TOPOLOGICAL INDICES OF GRAPHS

By: SAROJA Y. TALWAR

Copyright © [2023]

Title: **TOPOLOGICAL INDICES OF GRAPHS**

Author :   SAROJA Y. TALWAR

All rights reserved. No part of this publication may be reproduced, stored in a retrieval system, or transmitted in any form or by any means, electronic, mechanical, photocopying, recording, or otherwise, without the prior written permission of the publisher or author, except in the case of brief quotations embodied in critical reviews and certain other non-commercial uses permitted by copyright law.

This book was printed and published by

[Publisher's: **SAROJA Y. TALWAR**   in [2023]

ISBN:

For permission to reproduce any of the material in this book.

TABLE OF CONTENTS

1	Introduction	1
2	Reciprocal Status Connectivity Indices, Harmonic Reciprocal Status Index and Co-indices of Graphs	16
3	Status Connectivity Index, Reciprocal Status Connectivity Index and Harmonic Reciprocal Status Index of Line Graphs	59
4	Status Like Topological Indices of Graphs and its Regression Analysis with Some Molecular Properties	84
5	Zagreb Indices and Co-indices of Total graph, Semi-total Point Graph and Semi-total Line Graph of Subdivision Graphs	123
6	Friendship Network Analysis Using Status Connectivity Indices of Graphs	137
7	References	148

Chapter 1

Introduction

1.1 Preamble

Graph theory has become fashionable so much so that there are applications of graph theory in many areas like physics, chemistry, communication science, computer technology, electrical and civil engineering, architecture, operation research, genetics, psychology, sociology, economics, anthropology and linguistic. The theory is also intimately related to many branches of mathematics including group theory, matrix theory, numerical analysis, probability and topology. The fact is that graph theory serves as a mathematical model for any system involving a binary relation. Partly because of their diagrammatic representation, graphs have an intuitive and aesthetic appeal. Although there are many results in this field of an elementary nature, there is also an abundance of problems with enough combinatorial subtlety to challenge the most sophisticated mathematician.

In the recent years, the mathematical-chemistry literature is flooded by countless graph-based topological indices, proposed to serve as molecular structure descriptors. Topological indices have attracted much attention of chemical and mathematical researchers, especially those focusing on graph theory, from all over the world. Nowadays many interesting results and open problems on it have been reported in literature. In most cases, the mathematical investigation of these indices consist of finding bounds for them, and characterizing the graphs

Introduction

for which these inequalities become equalities.

1.2 Topological Indices

In the fields of chemical graph theory, molecular topology, and of mathematical chemistry, a topological index, is a type of a molecular descriptor, that is calculated based on the molecular graph of a chemical compound. Topological indices are numerical parameters of a graph which characterize its topology and are usually graph invariant [45]. Topological indices are used in the development of quantitative structure-activity relationships (QSAR) and in quantitative structure-property relationships (QSPR), in which the properties of molecules are correlated with their chemical structure.

QSAR/QSPR represents predictive models derived from application of statistical tools correlating activity (including desirable therapeutic effect and undesirable side effects) of chemicals (drugs / toxicants / environmental pollutants) with descriptors representative of molecular structure and/or properties [62].

Particular topological indices include the Harary index, Randic index, Zagreb index and Wiener index. Unless otherwise stated, hydrogen atoms are usually ignored in the computation of such indices as organic chemists usually do when they write a benzene ring as a hexagon.

Introduction

1.3 Definitions and Terminology

For definitions and terminologies of graphs we refer the books [29, 69]. A **graph** G consists of a finite nonempty set $V(G) = \{v_1, v_2, \ldots, v_n\}$ of n vertices together with a prescribed set $E(G)$ of m unordered pairs of distinct vertices of $V(G)$. Each pair of vertices in $E(G)$ is an edge of G. We consider only simple, undirected graphs without loops. Two vertices are said to be **adjacent** if there is an edge joining them. Then the edge between the vertices u and v is denoted by uv. An edge and its end vertex are **incident** with each other. Two edges are adjacent, if they are incident with common vertex.

The **degree** of a vertex u, denoted by $d_G(u)$ is the number of edges incident to it. If all the vertices have same degree equal to r, then G is called a **regular graph** of degree r or r-**regular graph**.

Two graphs G and H are said to be **isomorphic** if there is one-to-one correspondence between their vertex sets which preserves the adjacency, and it is written as $G \cong H$ or $G = H$.

A **walk** of a graph G is an alternating sequence of vertices and edges, beginning and ending with vertices, in which each edge is incident with the two vertices immediately preceding and following it. It is **closed** if end vertices are same and is **open**, otherwise. A walk is a **trail** if all the edges are distinct and it is a **path** if all the vertices are distinct. A path with n vertices is denoted by P_n. A closed walk

Introduction

is called the **cycle** provided all its n vertices are distinct and $n \geq 3$. A cycle with n vertices is denoted by C_n.

A graph is **connected** if every pair of vertices is joined by some path. The **length of a path** is the number of edges in it. The **distance** $d_G(u, v)$ between two vertices u and v of G is the length of shortest path joining u and v. The **diameter** of a connected graph G denoted by $D = diam(G)$, is the length of any longest geodesic.

The **eccentricity** $e_G(u)$ of a vertex u is the maximum distance between u and any other vertex of G.

The **status** [28] of a vertex $u \in V(G)$ is defined as the sum of its distance from every other vertex in $V(G)$ and is denoted by $\sigma_G(u)$. That is,

$$\sigma_G(u) = \sum_{v \in V(G)} d_G(u, v).$$

The **complement** $\overline{G}$ of a graph G is a graph with vertex set $V(G)$ and two vertices are adjacent in $\overline{G}$ if and only if they are not adjacent in G. A **self complementary graph** is isomorphic with its complement.

The **complete graph** K_n is a graph with n vertices in which every pair of vertices is adjacent. Thus K_n has $\frac{n(n-1)}{2}$ edges and it is regular graph of degree $n - 1$.

A **bipartite graph** is a graph whose vertex set can be partitioned into two subsets V_1 and V_2 such that no two vertices in either sets are adjacent. If every vertex of V_1 is adjacent to each vertex of V_2 then G is a **complete bipartite graph** and is denoted by $K_{p,q}$, where $|V_1| = p$ and $|V_2| = q$. A **star** on n vertices is a complete bipartite graph $K_{1,n-1}$.

The **line graph** of G, denoted by $L(G)$ is the graph whose vertices has one-to-one correspondence with the edges of G and two vertices of $L(G)$ are adjacent whenever the corresponding edges of G are adjacent. If G is an r-regular graph then $L(G)$ is $(2r - 2)$-regular graph.

A wheel W_{n+1} is a graph obtained from the cycle C_n by adding a new vertex and making it adjacent to all vertices of C_n.

A **friendship graph** (or **windmill graph**) F_n, $n \geq 2$, is a graph that can be constructed by coalescence of n copies of the cycle C_3 of length 3 with a common vertex. It has $2n + 1$ vertices and $3n$ edges. The degree of a coalescence vertex of F_n is $2n$ and the degree of all other vertices is 2. Also $diam(F_n) = 2$.

The **Kneser graph** $KG_{p,k}$ is the graph whose vertices correspond to the k-element subsets of a set of p elements, and where two vertices are adjacent if and only if the two corresponding sets are disjoint. Clearly we must impose the restriction $p \geq 2k$. The Kneser graph $KG_{p,k}$ has $\binom{p}{k}$ vertices and it is regular of degree $\binom{p-k}{k}$. Therefore the number of edges of $KG_{p,k}$ is $\frac{1}{2}\binom{p}{k}\binom{p-k}{k}$ (see [39]). The Kneser graph $KG_{n,1}$ is a complete graph on n vertices.

The vertices and edges of G are referred as their elements. The **total graph** of G, denoted by $T(G)$, is a graph with vertex set $V(T(G)) = V(G) \cup E(G)$ and two vertices in $T(G)$ are adjacent if and only if they are adjacent elements or incident elements in G [29].

The **semi-total point graph** of G, denoted by $T_1(G)$, is a graph with vertex set $V(T_1(G)) = V(G) \cup E(G)$ and two vertices in $T_1(G)$ are adjacent if they are adjacent vertices in G or one is vertex and other is an edge, incident to it [56].

The **semi-total line graph** of G, denoted by $T_2(G)$, is a graph with vertex set $V(T_2(G)) = V(G) \cup E(G)$ and two vertices in $T_2(G)$ are adjacent if they are adjacent edges in G or one is vertex and other is an edge, incident to it.

A **subdivision graph** is a graph which can be obtained from a given graph by inserting a new vertex on each edge of G.

The (m, n)-**tadpole graph**, also called a dragon graph, is the graph obtained by joining a cycle C_m to a path P_n with a bridge.

The **ladder graph** L_n is defined by $L_n = P_n \times K_2$ where P_n is a path with n vertices and $\times$ denotes the Cartesian product and K_2 is a complete graph with two vertices.

In the literature several degree based topological indices have been introduced and studied [24]. More studied topological indices based on the degree of vertices are Zagreb indices [23, 43]. The first and second **Zagreb indices** of a graph G are defined as [26],

$$Z_1(G) = \sum_{u \in V(G)} d_G(u)^2 = \sum_{uv \in E(G)} [d_G(u) + d_G(v)]. \qquad (1.1)$$

and

$$Z_2(G) = \sum_{uv \in E(G)} d_G(u)d_G(v).$$

The oldest distance based topological index is the **Wiener index** [70], defined as

$$W(G) = \sum_{\{u,v\} \subseteq V(G)} d_G(u, v) = \frac{1}{2} \sum_{u \in V(G)} \sigma_G(u).$$

The Wiener index is also called as gross status or total status [28].

The first and second **status connectivity indices** of a connected graph G are defined as [51]

$$S_1(G) = \sum_{uv \in E(G)} [\sigma_G(u) + \sigma_G(v)] \quad \text{and} \quad S_2(G) = \sum_{uv \in E(G)} \sigma_G(u)\sigma_G(v).$$

The **eccentric connectivity indices** of a connected graph G are defined as [4, 68]

$$\xi_1(G) = \sum_{uv \in E(G)} [e_G(u) + e_G(v)] \quad \text{and} \quad \xi_2(G) = \sum_{uv \in E(G)} [e(u)e(v)].$$

Details on mathematical properties and chemical applications of eccentric connectivity indices can be found in [5, 9, 15, 22, 31, 36, 41, 73].

The **Harary index** $HI(G)$ of a connected graph G is defined as the sum of reciprocal of the distances between all pairs of vertices of G [34, 45]. That is,

$$HI(G) = \sum_{\{u,v\} \subseteq V(G),\ u \neq v} \frac{1}{d_G(u, v)}.$$

The **Forgotten topological index** (also called as F-index) is defined as

$$F(G) = \sum_{v \in V(G)} d_G(v)^3 = \sum_{uv \in E(G)} \left(d_G(u)^2 + d_G(v)^2 \right).$$

The **sum-connectivity index** of a graph G, denoted by $SC(G)$, is defined as [74],

$$SC(G) = \sum_{uv \in E(G)} \frac{1}{\sqrt{d_G(u) + d_G(v)}}.$$

Estrada et al. [18] proposed a topological index called **atom-bond connectivity index**. It is defined as

$$ABC(G) = \sum_{uv \in E(G)} \sqrt{\frac{d_G(u) + d_G(v) - 2}{d_G(u) d_G(v)}}.$$

The **augmented Zagreb index** of a graph G, proposed by Furtula et al. [20], is defined as

$$AZ(G) = \sum_{uv \in E(G)} \left[\frac{d_G(u) d_G(v)}{d_G(u) + d_G(v) - 2} \right]^3.$$

Introduction

The **arithmetic-geometric index** of a graph G, proposed by Shigehalli and Kanabur [60], is defined as

$$AG(G) = \sum_{uv \in E(G)} \frac{d_G(u) + d_G(v)}{2\sqrt{d_G(u)d_G(v)}}.$$

The **geometric-arithmetic index** was invented by Vukicević and Furtula [67] and it is defined as

$$GA(G) = \sum_{uv \in E(G)} \frac{2\sqrt{d_G(u)d_G(v)}}{d_G(u) + d_G(v)}.$$

The **harmonic index** of a graph G is defined as [19]

$$H(G) = \sum_{uv \in E(G)} \frac{2}{d_G(u) + d_G(v)}.$$

Recent results on the harmonic index can be found in [7, 11, 30, 37, 38, 61, 72, 75].

The **harmonic status index** of a graph G is defined as [52]

$$HS(G) = \sum_{uv \in E(G)} \frac{2}{\sigma_G(u) + \sigma_G(v)}.$$

The original formulation of the Zagreb indices is presented and their relationship to topological indices such as self-returning walks, Platt, Gordon-Scantlebury and connectivity indices is discussed, along with their properties. Modified Zagreb indices are introduced and the Zagreb complexity indices reviewed. Their use in QSPR is illustrated by modeling the structure-boiling point relationship of alkanes. The obtained models are in fair agreement with experimental data and are

Introduction

better than many models in the literature. However, in general, the Zagreb indices do not contribute to the best structure-boiling point models of alkanes. Nevertheless, it is interesting to note that the best five-descriptor model that obtained in the literature contains the Zagreb index [43].

A formula to measure a positional aspect of the status of a person in an organization or a group, and investigate some of its ramifications. A recursion formula is derived which expresses the status of a person in a tree-organization in terms of the status numbers of his immediate subordinates. The index is then applied to the standard organization, to maximal and minimal status arrangements, the status pair, structural democracy, autocracy and laissez-faire, automorphic and peer groups. Problems with respect to the use of the formula and its relationship to other status formulas are discussed [28].

The boiling points of organic compounds, as well as all their physical properties, depend functionally upon the number, kind and structural arrangement of the atoms in the molecule. Within a group of isomers, both the number and the kind of atoms are constant, and variations in physical properties are due to changes in structural inter-relationships alone. The study of the effect of pure structural variation upon the boiling point of the paraffins may be expected to be of some theoretical interest and satisfactory results are readily obtained [70].

Some properties of the Schultz molecular topological index (MTI)

Introduction

are established which show why, in certain series of isomers, MTIs decreases with the increasing extent of branching of the molecular carbon-atom skeleton. The relation between MTI and the Wiener index is examined, and some of its hitherto unnoticed aspects are pointed out [23].

A new molecular graph matrix, the reciprocal distance (RD) matrix, and its non diagonal elements are equal to the reciprocals of the topological distances between the corresponding vertices, while the diagonal elements are all equal to zero. Based on the RD matrix, a real-number local vertex invariant, RDS was proposed. Their degeneracy was investigated and proved to be lower than that of the Wiener index based on the distance matrix. The correlational ability of the new molecular descriptors was tested against van der Waals molecular surfaces and boiling points of alkanes, showing a satisfactory monoparametric dependence in [34].

A novel topological index for the characterization of chemical graphs, derived from the reciprocal distance matrix and named the Harary index in honor of Professor Frank Harary, has been introduced. The Harary index is not a unique molecular descriptor; the smallest pair of the alkane trees with identical Harary indices has been detected in the octane family. The use of the Harary index in the quantitative structure-property relationships is exemplified in modeling physical properties of the alkanes. In this application, the performance of the

Introduction

Harary index is comparable to the performance of the Wiener index [45].

Extremal graphs with respect to Harary index, relationaship between Harary index and related topological indices, some properties and applications of Harary index are reported in [71].

The relationship of two graph invariants the eccentric connectivity index and the Wiener index was investigated with regard to anti-inflammatory activity, for a data set consisting of 76 pyrazole carboxylic acid hydrazide analogues. The values of the eccentric connectivity index and the Wiener index of each analogue in the data set were computed and active ranges were identified. Subsequently, each analogue was assigned a biological activity that was compared with the anti-inflammatory activity reported as percent reduction in paw swelling. Prediction with an accuracy of 90 percent was obtained using the eccentric connectivity index as compared to 84 percent in the case of Wiener index [22].

Das and Trinajstić [31] compared the eccentric connectivity index and Zagreb indices for chemical trees and molecular graphs. However, the comparison between the eccentric connectivity index and Zagreb indices, in the case of general trees and general graphs, is very hard and remains unsolved till now. The eccentric connectivity index and Zagreb indices for some graph families are compared. Then two classes of composite graphs, each of which has larger eccentric connectivity

Introduction

index than the first Zagreb index are introduced, if the original graph has larger eccentric connectivity index than the first Zagreb index. As a consequence, one can construct infinite classes of graphs having larger eccentric connectivity index than the first Zagreb index.

The general sum-connectivity index, general product-connectivity index, general Zagreb index and co-indices of line graphs of subdivision graphs of tadpole graphs, wheels and ladders have been reported in the literature and general expressions for these topological indices for the line graph of the subdivision graphs are reported [48].

1.4 Outline of the Present Work

The present thesis is divided into 6 chapters,

Chapter 1: This chapter is introductory in nature and deals with introduction of the thesis topic, bringing into consideration of some historical background, milestones lead by different researchers specially concentrating on chemical graph theory and hence recalling some fundamental definitions and terminologies.

Chapter 2: In this chapter, reciprocal status connectivity indices, Harmonic reciprocal status index and co-indices of graphs are introduced and studied. In the subsequent sections we compute the reciprocal status connectivity indices, Harmonic reciprocal status index and co-indices of some graphs and obtain the bounds for these indices.

Introduction

Further, we study the linear regression model between boiling point and reciprocal status connectivity indices of benzenoid hydrocarbons.

Chapter 3: In this chapter, the status connectivity index of line graphs of trees are obtained. Further we give the bounds for the status connectivity index of line graphs of connected graphs. Also obtain the results on first reciprocal status connectivity index and harmonic reciprocal status index of line graphs.

Chapter 4: In this chapter, we are dealing with topological indices based on the status and reciprocal status of vertices. Further we carry out the regression analysis with some molecular properties of paraffins such as melting point, boiling point, molecular mass and density.

Chapter 5: In this chapter we obtain the Zagreb indices and co-indices of the total graph of the subdivision graph of any graph. Thus generalizes the existing results. Further we compute the Zagreb indices of semi-total point graph and semi-total line graph of subdivision graph of any graph in terms of the parameters of underline graph.

Chapter 6: In this chapter the social network analysis by using first status connectivity index and second status connectivity index of graphs are discussed. The concept taken for study is how the connectivity increases in a peer group of students when they are brought together at one place for higher education.

Chapter 2

Reciprocal Status Connectivity Indices, Harmonic Reciprocal Status Index and Co-indices of Graphs

---

Results of this chapter are published in:

(i) H. S. Ramane, S. Y. Talwar, "Reciprocal status connectivity indices of graphs", Journal of Advanced Mathematical Studies, 12(3) (2019), 289-298. (Romania) [ISSN: 2065-3506].

(ii) H. S. Ramane, S. Y. Talwar, R Sharafdini, "Reciprocal status connectivity indices and co-indices of graphs", Indian Journal of Discrete Mathematics, 3(2) (2017), 61–72. (India) [ISSN: 2455-5819] .

(iii) H. S. Ramane, S. Y. Talwar "Harmonic reciprocal status index and co-index of graphs", Turkic World Mathematical Society Journal of Applied and Engineering Mathematics, (Accepted 2019). (Turkey) [ISSN: 2146-1147].

2.1 Introduction

The first and second *status connectivity indices* of a connected graph G are defined in [51].

We define here topological indices based on the reciprocal distances.

The *reciprocal status* of a vertex u is defined as the sum of reciprocal of its distances from every other vertex of G and is denoted by $rs_G(u)$. That is,

$$rs_G(u) = \sum_{v \in V(G),\, u \neq v} \frac{1}{d(u,v)}. \tag{2.1}$$

The *Harary index* $HI(G)$ of a connected graph G is defined as the sum of reciprocal of the distances between all pairs of vertices of G [45, 34]. That is,

$$HI(G) = \sum_{\{u,v\} \subseteq V(G),\, u \neq v} \frac{1}{d(u,v)} = \frac{1}{2} \sum_{u \in V(G)} rs_G(u). \tag{2.2}$$

Extremal graphs with respect to Harary index, relationaship between Harary index and related topological indices, some properties and applications of Harary index are reported in [71].

We define the *first reciprocal status connectivity index* $RS_1(G)$ and *second reciprocal status connectivity index* $RS_2(G)$ of a connected graph G as

$$RS_1(G) = \sum_{uv \in E(G)} [rs_G(u) + rs_G(v)] \qquad (2.3)$$

and

$$RS_2(G) = \sum_{uv \in E(G)} rs_G(u)rs_G(v). \qquad (2.4)$$

The *first reciprocal status connectivity co-index* $\overline{RS_1}(G)$ and *second reciprocal status connectivity co-index* $\overline{RS_2}(G)$ are defined as

$$\overline{RS_1}(G) = \sum_{uv \notin E(G)} [rs_G(u) + rs_G(v)] \qquad (2.5)$$

and

$$\overline{RS_2}(G) = \sum_{uv \notin E(G)} [rs_G(u)rs_G(v)]. \qquad (2.6)$$

The *harmonic index* of a graph G is defined in [19]. Recent results on the harmonic index can be found in [7, 11, 30, 37, 38, 61, 72, 75].

The *Harmonic status index* of a graph G is defined in [52].

Motivated by the harmonic index and harmonic status index of a graph, the harmonic reciprocal status index and harmonic reciprocal status co-index of connected graphs are introduced and studied.

The *harmonic reciprocal status index* of a connected graph G is defined as

$$HRS(G) = \sum_{uv \in E(G)} \frac{2}{rs_G(u) + rs_G(v)} \qquad (2.7)$$

and *harmonic reciprocal status co-index* of a connected graph G is defined as

$$\overline{HRS}(G) = \sum_{uv \notin E(G)} \frac{2}{rs_G(u) + rs_G(v)}. \tag{2.8}$$

For a graph given in Fig. 2.1, $RS_1(G) = 59$, $RS_2(G) \approx 108.25$, $\overline{RS_1}(G) = 13$, $\overline{RS_2}(G) = 21$, $HRS(G) \approx 2.1738$ and $\overline{HRS}(G) \approx 0.6153$.

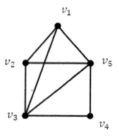

Figure 2.1 : Graph G

From Eq. (2.2) we observe that the Harary index has a relation with reciprocal status of a vertices. Although the several results have been obtained on Harary index (see [45, 48, 71]), it will be interesting to study further properties of the discrete structure using reciprocal status connectivity indices and co-indices.

In the subsequent sections of this chapter we compute the reciprocal status connectivity indices, harmonic reciprocal status indices and co-indices of some graphs and obtain the bounds for these indices. Further, we study the linear regression model between boiling point and reciprocal status connectivity indices of benzenoid hydrocarbons. Also obtain reciprocal status connectivity indices and co-indices of some reciprocal status regular graphs.

2.2 Computation of Reciprocal Status Connectivity Indices

Theorem 2.2.1. *Let G be a connected graph on n vertices and m edges. Let $diam(G) \leq 2$. Then*

$$RS_1(G) = m(n-1) + \frac{1}{2}Z_1(G)$$

and

$$RS_2(G) = \frac{1}{4}\left[m(n-1)^2 + (n-1)Z_1(G) + Z_2(G)\right].$$

Proof. If $diam(G) \leq 2$, then $d_G(u)$ number of vertices are at distance 1 from the vertex u and the remaining $n - 1 - d_G(u)$ vertices are at distance 2. Therefore for each vertex u in G, $rs_G(u) = \frac{1}{2}[n-1+d_G(u)]$. Therefore,

$$
\begin{aligned}
RS_1(G) &= \sum_{uv \in E(G)} [rs_G(u) + rs_G(v)] \\
&= \sum_{uv \in E(G)} \frac{1}{2}[n - 1 + d_G(u) + n - 1 + d_G(v)] \\
&= \sum_{uv \in E(G)} (n - 1) + \frac{1}{2} \sum_{uv \in E(G)} [d_G(u) + d_G(v)] \\
&= m(n - 1) + \frac{1}{2}Z_1(G)
\end{aligned}
$$

and

$$RS_2(G) = \sum_{uv \in E(G)} rs_G(u)rs_G(v)$$

$$= \sum_{uv \in E(G)} \frac{1}{4}(n - 1 + d_G(u))(n - 1 + d_G(v))$$

$$= \frac{1}{4}\left[m(n-1)^2 + (n-1)\sum_{uv \in E(G)} [d_G(u) + d_G(v)] \right.$$

$$\left. + \sum_{uv \in E(G)} d_G(u)d_G(v) \right]$$

$$= \frac{1}{4}\left[m(n-1)^2 + (n-1)Z_1(G) + Z_2(G) \right].$$

$\square$

Corollary 2.2.2. *Let G be a connected regular graph of degree r on n vertices and m edges. Let $diam(G) \leq 2$. Then*

$$RS_1(G) = m(n-1) + mr$$

and

$$RS_2(G) = \frac{1}{4}\left[m(n-1)^2 + 2mr(n-1) + mr^2 \right].$$

Corollary 2.2.3. *For a complete graph K_n on n vertices,*

$$RS_1(K_n) = n(n-1)^2 \ \text{ and } \ RS_2(K_n) = \frac{1}{2}n(n-1)^3.$$

By Theorem 2.2.1, we have the following result.

Proposition 2.2.4. *For a complete bipartite graph $K_{p,q}$,*

$$RS_1(K_{p,q}) = \frac{1}{2}pq\left[3(p+q) - 2 \right]$$

and

$$RS_2(K_{p,q}) = \frac{pq}{4}\left[(p+q-1)(2p+2q-1)+pq\right].$$

Proposition 2.2.5. *For a path P_n on n vertices,*

$$RS_1(P_n) = 2\sum_{j=1}^{n-1}\frac{1}{j} + 2\sum_{i=2}^{n-1}\left[\sum_{j=1}^{i-1}\frac{1}{j} + \sum_{j=1}^{n-i}\frac{1}{j}\right]$$

and

$$\begin{aligned}
RS_2(P_n) &= 2\left(\sum_{j=1}^{n-1}\frac{1}{j}\right)\left(1+\sum_{j=1}^{n-2}\frac{1}{j}\right) \\
&\quad + \sum_{i=2}^{n-2}\left[\left(\sum_{j=1}^{i-1}\frac{1}{j} + \sum_{j=1}^{n-i}\frac{1}{j}\right)\left(\sum_{j=1}^{i}\frac{1}{j} + \sum_{j=1}^{n-i-1}\frac{1}{j}\right)\right].
\end{aligned}$$

Proof. Let $v_1, v_2, \ldots, v_n$ be the vertices of the path P_n, where v_i is adjacent to v_{i+1}, $i = 1, 2, \ldots, n-1$. Therefore,

$$rs_{P_n}(v_1) = rs_{P_n}(v_n) = \sum_{j=1}^{n-1}\frac{1}{j}$$

and

$$rs_{P_n}(v_i) = \sum_{j=1}^{i-1}\frac{1}{j} + \sum_{j=1}^{n-i}\frac{1}{j}, \quad \text{for} \ \ i = 2, 3, \ldots, n-1.$$

Therefore

$$\begin{aligned}
RS_1(P_n) &= \sum_{i=1}^{n-1}\left[rs_{P_n}(v_i) + rs_{P_n}(v_{i+1})\right] \\
&= \left[rs_{P_n}(v_1) + rs_{P_n}(v_n)\right] + 2\sum_{i=2}^{n-1}rs_{P_n}(v_i) \\
&= 2\sum_{j=1}^{n-1}\frac{1}{j} + 2\sum_{i=2}^{n-1}\left[\sum_{j=1}^{i-1}\frac{1}{j} + \sum_{j=1}^{n-i}\frac{1}{j}\right]
\end{aligned}$$

and

$$
\begin{aligned}
RS_2(P_n) &= \sum_{i=1}^{n-1} rs_{P_n}(v_i)rs_{P_n}(v_{i+1}) \\
&= rs_{P_n}(v_1)rs_{P_n}(v_2) \\
&\quad + \sum_{i=2}^{n-2} rs_{P_n}(v_i)rs_{P_n}(v_{i+1}) + rs_{P_n}(v_{n-1})rs_{P_n}(v_n) \\
&= \left(\sum_{j=1}^{n-1}\frac{1}{j}\right)\left(1+\sum_{j=1}^{n-2}\frac{1}{j}\right) \\
&\quad + \sum_{i=2}^{n-2}\left[\left(\sum_{j=1}^{i-1}\frac{1}{j}+\sum_{j=1}^{n-i}\frac{1}{j}\right)\left(\sum_{j=1}^{i}\frac{1}{j}+\sum_{j=1}^{n-i-1}\frac{1}{j}\right)\right] \\
&\quad + \left(1+\sum_{j=1}^{n-2}\frac{1}{j}\right)\left(\sum_{j=1}^{n-1}\frac{1}{j}\right) \\
&= 2\left(\sum_{j=1}^{n-1}\frac{1}{j}\right)\left(1+\sum_{j=1}^{n-2}\frac{1}{j}\right) \\
&\quad + \sum_{i=2}^{n-2}\left[\left(\sum_{j=1}^{i-1}\frac{1}{j}+\sum_{j=1}^{n-i}\frac{1}{j}\right)\left(\sum_{j=1}^{i}\frac{1}{j}+\sum_{j=1}^{n-i-1}\frac{1}{j}\right)\right].
\end{aligned}
$$

$\square$

Proposition 2.2.6. *For a cycle C_n on $n \geq 3$ vertices,*

$$
RS_1(C_n) = \begin{cases}
4 + 4n\sum_{i=1}^{\frac{n-2}{2}}\frac{1}{i} & \text{if } n \text{ is even} \\[2em]
4n\sum_{i=1}^{\frac{n-1}{2}}\frac{1}{i} & \text{if } n \text{ is odd}
\end{cases}
$$

and

$$RS_2(C_n) = \begin{cases} n\left[\frac{2}{n} + 2\sum_{i=1}^{\frac{n-2}{2}}\frac{1}{i}\right]^2 & \text{if } n \text{ is even} \\ 4n\left[\sum_{i=1}^{\frac{n-1}{2}}\frac{1}{i}\right]^2 & \text{if } n \text{ is odd} \end{cases}.$$

Proof. If n is even number, then for every vertex u of C_n,

$$rs_{C_n}(u) = \frac{2}{n} + 2\sum_{i=1}^{\frac{n-2}{2}}\frac{1}{i}.$$

Therefore

$$\begin{aligned} RS_1(C_n) &= \sum_{uv\in E(C_n)} \left[rs_{C_n}(u) + rs_{C_n}(v)\right] \\ &= \sum_{uv\in E(C_n)} \left[\frac{4}{n} + 4\sum_{i=1}^{\frac{n-2}{2}}\frac{1}{i}\right] \\ &= 4 + 4n\sum_{i=1}^{\frac{n-2}{2}}\frac{1}{i} \end{aligned}$$

and

$$\begin{aligned} RS_2(C_n) &= \sum_{uv\in E(C_n)} rs_{C_n}(u)rs_{C_n}(v) \\ &= \sum_{uv\in E(C_n)} \left[\frac{2}{n} + 2\sum_{i=1}^{\frac{n-2}{2}}\frac{1}{i}\right]^2 \\ &= n\left[\frac{2}{n} + 2\sum_{i=1}^{\frac{n-2}{2}}\frac{1}{i}\right]^2. \end{aligned}$$

If n is odd number, then for every vertex u of C_n,

$$rs_{C_n}(u) = 2\sum_{i=1}^{\frac{n-1}{2}}\frac{1}{i}.$$

Therefore

$$RS_1(C_n) = \sum_{uv \in E(C_n)} [rs_{C_n}(u) + rs_{C_n}(v)]$$

$$= \sum_{uv \in E(C_n)} 4 \sum_{i=1}^{\frac{n-1}{2}} \frac{1}{i}$$

$$= 4n \sum_{i=1}^{\frac{n-1}{2}} \frac{1}{i}$$

and

$$RS_2(C_n) = \sum_{uv \in E(C_n)} rs_{C_n}(u) rs_{C_n}(v)$$

$$= \sum_{uv \in E(C_n)} \left(2 \sum_{i=1}^{\frac{n-1}{2}} \frac{1}{i} \right)^2$$

$$= 4n \left[\sum_{i=1}^{\frac{n-1}{2}} \frac{1}{i} \right]^2 .$$

$\square$

The wheel W_{n+1} has $n+1$ vertices and $2n$ edges. Also $diam(W_{n+1}) \leq 2$ and $Z_1(W_{n+1}) = n^2 + 9n$ and $Z_2(W_{n+1}) = 3n^2 + 9n$. Substituting these in Theorem 2.2.1 we get the following results.

Proposition 2.2.7. *For a wheel $W_{n+1}, n \geq 3$,*

$$RS_1(W_{n+1}) = \frac{n}{2}(5n + 9) \text{ and } RS_2(W_{n+1}) = \frac{1}{4} \left[3n^3 + 12n^2 + 9n \right].$$

A *friendship graph* (or *windmill graph*) F_n, has $2n+1$ vertices and $3n$ edges. The degree of a coalescence vertex of F_n is $2n$ and the degree of all other vertices is 2. Also $diam(F_n) = 2$ and $Z_1(F_n) = 4n^2 + 8n$ and $Z_2(F_n) = 8n^2 + 4n$. Therefore by Theorem 2.2.1, we have following result.

Proposition 2.2.8. *For a friendship graph F_n, $n \geq 2$,*

$$RS_1(F_n) = 4n(2n+1) \ \ and \ \ RS_2(F_n) = 5n^3 + 6n^2 + n.$$

Proposition 2.2.9. *If G is a connected regular graph of degree r, then $RS_1(G) = 2rHI(G)$.*

Proof.

$$
\begin{aligned}
RS_1(G) &= \sum_{uv \in E(G)} [rs_G(u) + rs_G(v)] \\
&= \sum_{u \in V(G)} rs_G(u)d_G(u) \\
&= r \sum_{u \in V(G)} rs_G(u) \\
&= 2rHI(G).
\end{aligned}
$$

$\square$

2.3 Bounds for the Reciprocal Status Connectivity Indices

Theorem 2.3.1. *Let G be a connected, nontrivial graph on n vertices and m edges and let $diam(G) = D$. Then*

$$\frac{2m}{D}(n-1) + \left(1 - \frac{1}{D}\right) Z_1(G) \leq RS_1(G) \leq m(n-1) + \frac{1}{2}Z_1(G)$$

and

$$m \left(\tfrac{n-1}{D}\right)^2 + \left(\tfrac{n-1}{D}\right)\left(1 - \tfrac{1}{D}\right) Z_1(G) + \left(1 - \tfrac{1}{D}\right)^2 Z_2(G)$$

$$\leq RS_2(G)$$

$$\leq \tfrac{1}{4}[m(n-1)^2 + (n-1)Z_1(G) + Z_2(G)].$$

Equality in both cases holds if and only if $diam(G) \leq 2$.

Proof. <u>Upper bound</u>: For any vertex u of G, there are $d_G(u)$ vertices which are at distance 1 from u and remaining $n - 1 - d_G(u)$ vertices are at distance at least 2. Therefore

$$rs_G(u) \leq d_G(u) + \frac{1}{2}(n - 1 - d_G(u)) = \frac{1}{2}[n - 1 + d_G(u)].$$

Therefore

$$\begin{aligned} RS_1(G) &= \sum_{uv \in E(G)} [rs_G(u) + rs_G(v)] \\ &\leq \sum_{uv \in E(G)} \left[(n-1) + \frac{1}{2}(d_G(u) + d_G(v)) \right] \\ &= m(n-1) + \frac{1}{2}Z_1(G) \end{aligned}$$

and

$$\begin{aligned} RS_2(G) &= \sum_{uv \in E(G)} rs_G(u)rs_G(v) \\ &\leq \sum_{uv \in E(G)} \frac{1}{4}\left[(n-1)^2 + (n-1)(d_G(u) + d_G(v)) + d_G(u)d_G(v) \right] \\ &= \frac{1}{4}\left[m(n-1)^2 + (n-1)Z_1(G) + Z_2(G) \right]. \end{aligned}$$

<u>Lower bound:</u> For any vertex u of G, there are $d_G(u)$ vertices which are at distance 1 from u and the remaining $n-1-d_G(u)$ vertices are at distance at most D. Therefore

$$rs_G(u) \geq d_G(u) + \frac{1}{D}(n-1-d_G(u)) = \frac{1}{D}(n-1) + \left(1 - \frac{1}{D}\right)d_G(u).$$

Therefore

$$
\begin{aligned}
RS_1(G) &= \sum_{uv \in E(G)} [rs_G(u) + rs_G(v)] \\
&\geq \sum_{uv \in E(G)} \left[\frac{2}{D}(n-1) + \left(1 - \frac{1}{D}\right)(d_G(u) + d_G(v)) \right] \\
&= \frac{2m}{D}(n-1) + \left(1 - \frac{1}{D}\right)Z_1(G)
\end{aligned}
$$

and

$$
\begin{aligned}
RS_2(G) &= \sum_{uv \in E(G)} rs_G(u)rs_G(v) \\
&\geq \sum_{uv \in E(G)} \left(\frac{n-1}{D}\right)^2 + \left(\frac{n-1}{D}\right)\left(1 - \frac{1}{D}\right)(d_G(u) + d_G(v)) \\
&\quad + \left(1 - \frac{1}{D}\right)^2 d_G(u)d_G(v) \\
&= m\left(\frac{n-1}{D}\right)^2 + \left(\frac{n-1}{D}\right)\left(1 - \frac{1}{D}\right)Z_1(G) \\
&\quad + \left(1 - \frac{1}{D}\right)^2 Z_2(G).
\end{aligned}
$$

For equality: If the diameter of G is 1 or 2 then the equality holds.

Conversely, let $RS_1(G) = m(n-1) + \frac{1}{2}Z_1(G)$.

Suppose, $diam(G) \geq 3$, then there exists at least one pair of vertices, say u_1 and u_2 such that $d_G(u_1, u_2) \geq 3$.

Therefore $rs_G(u_1) \leq d_G(u_1) + \frac{1}{3} + \frac{1}{2}(n - 2 - d_G(u_1)) = \frac{n}{2} - \frac{2}{3} + \frac{d_G(u_1)}{2}$.

Similarly $rs_G(u_2) \leq \frac{n}{2} - \frac{2}{3} + \frac{d_G(u_2)}{2}$ and for all other vertices u of G,

$rs_G(u) \leq \frac{n}{2} - \frac{1}{2} + \frac{d_G(u)}{2}$.

Partition the edge set of G into three sets E_1, E_2 and E_3, such that

$$E_1 = \left\{ u_1 v \mid rs_G(u_1) \leq \frac{n}{2} - \frac{2}{3} + \frac{d_G(u_1)}{2} \text{ and } rs_G(v) \leq \frac{n}{2} - \frac{1}{2} + \frac{d_G(v)}{2} \right\},$$

$$E_2 = \left\{ u_2 v \mid rs_G(u_2) \leq \frac{n}{2} - \frac{2}{3} + \frac{d_G(u_2)}{2} \text{ and } rs_G(v) \leq \frac{n}{2} - \frac{1}{2} + \frac{d_G(v)}{2} \right\}$$

and

$$E_3 = \left\{ uv \mid rs_G(u) \leq \frac{n}{2} - \frac{1}{2} + \frac{d_G(u)}{2} \text{ and } rs_G(v) \leq \frac{n}{2} - \frac{1}{2} + \frac{d_G(v)}{2} \right\}.$$

Easily we check that $|E_1| = d_G(u_1)$, $|E_2| = d_G(u_2)$ and $|E_3| = m - d_G(u_1) - d_G(u_2)$. Therefore

$$
\begin{aligned}
RS_1(G) &= \sum_{uv \in E(G)} [rs_G(u) + rs_G(v)] \\
&= \sum_{u_1 v \in E_1} [rs_G(u_1) + rs_G(v)] + \sum_{u_2 v \in E_2} [rs_G(u_2) + rs_G(v)] \\
&\quad + \sum_{uv \in E_3} [rs_G(u) + rs_G(v)]
\end{aligned}
$$

$$\geq \sum_{u_1v\in E_1} \left[\left(n-\frac{7}{6}\right)+\frac{1}{2}(d_G(u_1)+d_G(v))\right]$$

$$+\sum_{u_2v\in E_2} \left[\left(n-\frac{7}{6}\right)+\frac{1}{2}(d_G(u_2)+d_G(v))\right]$$

$$+\sum_{uv\in E_3} \left[(n-1)+\frac{1}{2}(d_G(u)+d_G(v))\right]$$

$$=\left(n-\frac{7}{6}\right)d_G(u_1)+\left(n-\frac{7}{6}\right)d_G(u_2)$$

$$+(n-1)\left[m-d_G(u_1)-d_G(u_2)\right]$$

$$+\frac{1}{2}\sum_{uv\in E(G)}[d_G(u)+d_G(v)]$$

$$=m(n-1)-\frac{1}{6}[d_G(u_1)+d_G(u_2)]+\frac{1}{2}Z_1(G),$$

which is a contradiction. Hence $diam(G)\leq 2$.

Similarly we can prove the equality of $RS_2(G)$. $\square$

Corollary 2.3.2. *Let G be a connected, nontrivial graph on n vertices and m edges and let $diam(G)=D$. Let δ and Δ be the minimum and maximum degree of the vertices of G repectively. Then*

$$\frac{2m}{D}(n-1)+\left(1-\frac{1}{D}\right)2m\delta \leq RS_1(G)\leq m(n-1)+m\Delta$$

and

$$m\left(\frac{n-1}{D}\right)^2+\left(\frac{n-1}{D}\right)\left(1-\frac{1}{D}\right)2m\delta+\left(1-\frac{1}{D}\right)^2m\delta^2$$

$$\leq RS_2(G)$$

$$\leq \tfrac{1}{4}[m(n-1)^2+2m(n-1)\Delta+m\Delta^2].$$

Proof. For any vertex u of G, $\delta\leq d_G(u)\leq \Delta$.

Therefore $2m\delta \leq Z_1(G) \leq 2m\Delta$ and $m\delta^2 \leq Z_2(G) \leq m\Delta^2$. Substituting these in Theorem 2.3.1, we get the results. $\square$

Corollary 2.3.3. *Let G be a connected regular graph of degree r on n vertices and m edges. Let $diam(G) = D$. Then*

$$\frac{2m}{D}(n-1) + \left(1 - \frac{1}{D}\right) 2mr \leq RS_1(G) \leq m(n-1) + mr$$

and

$$m \left(\tfrac{n-1}{D}\right)^2 + \left(\tfrac{n-1}{D}\right) \left(1 - \tfrac{1}{D}\right) 2mr + \left(1 - \tfrac{1}{D}\right)^2 mr^2$$

$$\leq RS_2(G)$$

$$\leq \tfrac{1}{4}[m(n-1)^2 + 2mr(n-1) + mr^2].$$

Equality in both cases holds if and only if $diam(G) \leq 2$.

Proof. If G is a regular graph, then $Z_1(G) = 2mr$ and $Z_2(G) = mr^2$. Therefore the result follows from the Theorem 2.3.1. $\square$

2.4 Regression Model

In this section the correlation between the boiling point (BP) of benzenoid hydrocarbons and the distance based indices of the corresponding molecular graphs is discussed. The correlation of boiling point with status connectivity indices, eccentric connectivity indices and Wiener index of benzenoid hydrocarbons given in Fig. 2.2 have been considered in [51]. The correlation coefficient between boiling point and status connectivity indices, eccentric connectivity indices and Wiener index of benzenoid hydrocarbons are given in Table 2.1 .

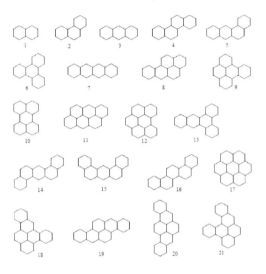

Figure 2.2 : Molecular graphs of benzenoid hydrocarbons under consideration.

The linear regression models for the boiling point (BP) of benzenoid hydrocarbons with status connectivity indices, eccentric connectivity indices and Wiener index are as below [51]

$$BP = 255.612(\pm 15.098) + 0.06(\pm 0.004)S_1 \qquad (2.9)$$

$$BP = 329.082(\pm 18.507) + 0.001(\pm 0.0)S_2 \qquad (2.10)$$

$$BP = 199.578(\pm 28.269) + 0.899(\pm 0.084)\xi_1 \qquad (2.11)$$

$$BP = 291.549(\pm 33.537) + 0.172(\pm 0.027)\xi_2 \qquad (2.12)$$

$$BP = 266.263(\pm 26.051) + 0.308(\pm 0.033)W \qquad (2.13)$$

Table 2.1: Correlation coefficient and standard error of the estimation [51]

Index	Correlation Coefficient (R) with boiling point	Standard error of the estimate
S_1	0.968	25.866
S_2	0.916	41.206
ξ_1	0.927	38.525
ξ_2	0.826	57.848
W	0.904	43.815

Here we discuss the correlation between the boiling point (BP) and reciprocal status connectivity indices and Harary index of benzenoid hydrocarbons given in Fig. 2.2.

Experimental values of boiling points of benzenoid hydrocarbons represented in Fig. 2.2 are taken from [44].

The scatter plot between BP and indices RS_1, RS_2 and H are shown in Figs. 2.3 to 2.5.

The linear regression models for the boiling point (BP) using the data of Table 2.2 are obtained by the least square fitting procedure as implemented in SPSS (Statistical Package for the Social Sciences) Statistics programme, and they are:

$$BP = 185.735(\pm 20.033) + 0.848(\pm 0.053)RS_1 \qquad (2.14)$$

$$BP = 239.887(\pm 15.376) + 0.176(\pm 0.010)RS_2 \qquad (2.15)$$

$$BP = 139.056(\pm 11.975) + 4.716(\pm 0.155)H \qquad (2.16)$$

Table 2.2: The values of experimental boiling points, reciprocal status connectivity indices and of Harary index of 21 benzenoid hydrocarbons.

Benzenoid hydrocarbon	BP in ^{0}C	RS_1	RS_2	H
1	218	106.93	260.86	23.89
2	338	192.13	580.90	41.14
3	340	189.94	567.41	40.78
4	431	223.52	1007.47	60.70
5	425	287.46	992.10	60.36
6	429	295.59	1051.67	61.72
7	440	284.29	969.72	59.83
8	496	365.94	1406.34	74.37
9	493	366.82	1414.48	74.54
10	497	366.41	1410.33	74.48
11	547	437.95	1789.65	87.31
12	542	441.05	1815.46	87.88
13	535	403.18	1581.61	83.25
14	536	333.90	1504.38	81.56
15	531	382.12	1472.53	81.93
16	519	396.43	1529.31	82.06
17	590	522.77	2293.26	102.34
18	592	485.21	2054.12	97.91
19	596	477.43	1986.24	96.59
20	594	477.23	1984.67	90.54
21	595	487.74	2074.72	98.35

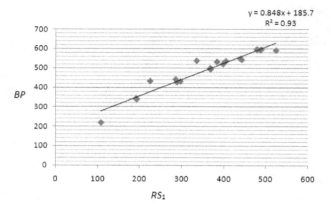

Figure 2.3: Scatter plot between the boiling point (BP) and the first reciprocal status connectivity index (RS_1).

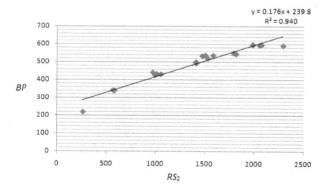

Figure 2.4: Scatter plot between the boiling point (BP) and the second reciprocal status connectivity index (RS_2).

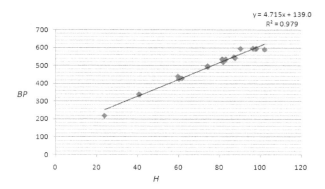

Figure 2.5: Scatter plot between the boiling point (BP) and the Harary index (HI).

Table 2.3: Correlation coefficient and standard error of the estimation

Index	Correlation Coefficient (R) with boiling point	Standard error of the estimate
RS_1	0.964	27.132
RS_2	0.970	24.949
HI	0.990	14.565

From Table 2.2 and Table 2.3, we observe that the models (2.14) (2.15) and (2.16) show that the correlation of the experimental boiling point of benzenoid hydrocarbons with reciprocal status connectivity indices and with Harary index is better ($R = 0.964, 0.970, 0.990$ respectively) than the correlation with other distance based indices like status connectivity indices, eccentric connectivity indices and Wiener index.

2.5 Reciprocal Status Connectivity Co-indices of Graphs

Proposition 2.5.1. *Let G be a connected graph on n vertices. Then*

$$\overline{RS_1}(G) = 2(n-1)HI(G) - RS_1(G) \qquad (2.17)$$

and

$$\overline{RS_2}(G) = 2(HI(G))^2 - \frac{1}{2}\sum_{u\in V(G)}(rs_G(u))^2 - RS_2(G). \qquad (2.18)$$

Proof.

$$
\begin{aligned}
\overline{RS_1}(G) &= \sum_{uv\notin E(G)}[rs_G(u) + rs_G(v)] \\
&= \sum_{\{u,v\}\subseteq V(G)}[rs_G(u) + rs_G(v)] - \sum_{uv\in E(G)}[rs_G(u) + rs_G(v)] \\
&= (n-1)\sum_{u\in V(G)}rs_G(u) - RS_1(G) \\
&= 2(n-1)HI(G) - RS_1(G)
\end{aligned}
$$

and

$$
\begin{aligned}
\overline{RS_2}(G) &= \sum_{un\notin E(G)}[rs_G(u)rs_G(v)] \\
&= \sum_{\{u,v\}\subseteq V(G)}[rs_G(u)rs_G(v)] - \sum_{uv\in E(G)}[rs_G(u)rs_G(v)] \\
&= \frac{1}{2}\left[\left(\sum_{u\in V(G)}rs_G(u)\right)^2 - \sum_{u\in V(G)}(rs_G(u))^2\right] - RS_2(G) \\
&= 2(HI(G))^2 - \frac{1}{2}\sum_{u\in V(G)}(rs_G(u))^2 - RS_2(G).
\end{aligned}
$$

$\square$

Corollary 2.5.2. *Let G be a connected graph with n vertices, m edges and $diam(G) \leq 2$. Then*

$$\overline{RS_1}(G) = 2(n-1)\left[\frac{1}{4}(2m + n(n-1))\right] - \left[(n-1)m + \frac{1}{2}Z_1(G)\right]$$
$$(2.19)$$

and

$$\overline{RS_2}(G) = \frac{1}{8}(2m + n(n-1))^2 - \frac{1}{8}\left[n(n-1)^2 + 4m(n-1) + Z_1(G)\right]$$
$$- \frac{1}{4}\left[(n-1)^2 m + (n-1)Z_1(G) + Z_2(G)\right]. \qquad (2.20)$$

Proof. For any graph G of $diam(G) \leq 2$, $rs_G(u) = \frac{1}{2}[(n-1) + d_G(u)]$ and

$$HI(G) = m + \frac{1}{2}\left[\frac{n(n-1)}{2} - m\right] = \frac{1}{4}[2m + n(n-1)].$$

Also we have

$$RS_1(G) = (n-1)m + \frac{1}{2}Z_1(G) \quad \text{and}$$
$$RS_2(G) = \frac{1}{4}\left[(n-1)^2 m + (n-1)Z_1(G) + Z_2(G)\right].$$

Therefore by the Proposition 2.5.1,

$$\overline{RS_1}(G) = 2(n-1)HI(G) - RS_1(G)$$
$$= 2(n-1)\left[\frac{1}{4}(2m + n(n-1))\right] - \left[(n-1)m + \frac{1}{2}Z_1(G)\right]$$

and

$$\overline{RS_2}(G) = 2(HI(G))^2 - \frac{1}{2}\sum_{u \in V(G)}(rs_G(u))^2 - RS_2(G)$$

$$= 2\left[\frac{1}{4}(2m + n(n-1))\right]^2 - \frac{1}{2}\sum_{u \in V(G)}(rs_G(u))^2$$
$$- \left(\frac{1}{4}\left[(n-1)^2 m + (n-1)Z_1(G) + Z_2(G)\right]\right)$$

$$= 2\left[\frac{1}{4}(2m + n(n-1))\right]^2 - \frac{1}{2}\sum_{u \in V(G)}\left[\frac{1}{2}((n-1) + d_G(u))\right]^2$$
$$- \left[\frac{1}{4}((n-1)^2 m + (n-1)Z_1(G) + Z_2(G))\right]$$

$$= 2\left[\frac{1}{4}(2m + n(n-1))\right]^2 - \frac{1}{2}\sum_{u \in V(G)}\left(\frac{1}{2}[(n-1) + d_G(u)]\right)^2$$
$$- \left[\frac{1}{4}((n-1)^2 m + (n-1)Z_1(G) + Z_2(G))\right]$$

$$= 2\left[\frac{1}{4}(2m + n(n-1))\right]^2 - \frac{1}{8}\left[\sum_{u \in V(G)}(n-1)^2 + 2(n-1)\sum_{u \in V(G)}d_G(u)\right.$$
$$\left. + \sum_{u \in V(G)}d_G(u)^2\right] - RS_2(G)$$

$$= \frac{1}{8}(2m + n(n-1))^2 - \frac{1}{8}\left[n(n-1)^2 + 4m(n-1) + Z_1(G)\right]$$
$$- \frac{1}{4}\left[(n-1)^2 m + (n-1)Z_1(G) + Z_2(G)\right].$$

$\square$

Proposition 2.5.3. *Let G be a connected graph with n vertices, m*

edges and $diam(G) \leq 2$. Then

$$\overline{RS_1}(G) = \frac{1}{2}n(n-1)^2 - m(n-1) + \frac{1}{2}\overline{Z_1}(G) \qquad (2.21)$$

and

$$\overline{RS_2}(G) = \frac{1}{4}\left[\frac{1}{2}n(n-1)^3 - m(n-1)^2 + (n-1)\overline{Z_1}(G) + \overline{Z_2}(G)\right].$$
$$(2.22)$$

Proof. For any graph G of $diam(G) \leq 2$, $rs_G(u) = \frac{1}{2}[(n-1) + d_G(u)]$.
Therefore

$$
\begin{aligned}
\overline{RS_1}(G) &= \sum_{uv \notin E(G)} \left[\left[\frac{1}{2}((n-1) + d_G(u))\right] + \left[\frac{1}{2}((n-1) + d_G(v))\right]\right] \\
&= \sum_{uv \notin E(G)} \left[(n-1) + \frac{1}{2}(d_G(u) + d_G(v))\right] \\
&= (n-1) \sum_{uv \notin E(G)} 1 + \frac{1}{2} \sum_{uv \notin E(G)} [d_G(u) + d_G(v)] \\
&= (n-1) \left(\frac{n(n-1)}{2} - m\right) + \frac{1}{2}\overline{Z_1}(G) \\
&= \frac{1}{2}n(n-1)^2 - m(n-1) + \frac{1}{2}\overline{Z_1}(G)
\end{aligned}
$$

and

$$
\begin{aligned}
\overline{RS_2}(G) &= \sum_{uv \notin E(G)} [rs_G(u)rs_G(v)] \\
&= \sum_{uv \notin E(G)} \frac{1}{4}((n-1) + d_G(u))((n-1) + d_G(v))
\end{aligned}
$$

$$= \sum_{uv \notin E(G)} \frac{1}{4} \left[(n-1)^2 + d_G(u)d_G(v) + (n-1)(d_G(u) + d_G(v)) \right]$$

$$= \frac{1}{4} \left[(n-1)^2 \sum_{uv \notin E(G)} 1 + \sum_{uv \notin E(G)} d_G(u)d_G(v) \right.$$

$$\left. + (n-1) \sum_{uv \notin E(G)} [d_G(u) + d_G(v)] \right]$$

$$= \frac{1}{4} \left[(n-1)^2 \left(\frac{n(n-1)}{2} - m \right) + \overline{Z_2}(G) + (n-1)\overline{Z_1}(G) \right]$$

$$= \frac{1}{4} \left[\frac{1}{2} n(n-1)^3 - m(n-1)^2 + (n-1)\overline{Z_1}(G) + \overline{Z_2}(G) \right].$$

$\square$

Proposition 2.5.4. *Let G be a graph with n vertices and m edges. Let $\overline{G}$, the complement of G, be connected. Then*

$$RS_1(\overline{G}) \le n(n-1)^2 - 2m(n-1) - \frac{1}{2}\overline{Z_1}(G) \qquad (2.23)$$

and

$$RS_2(\overline{G}) \le (n-1)^2 \left(\frac{n(n-1)}{2} - m \right) - \frac{1}{2}(n-1)\overline{Z_1}(G) + \frac{1}{4}\overline{Z_2}(G). \quad (2.24)$$

Proof. For any vertex u in $\overline{G}$ there are $n - 1 - d_G(u)$ vertices which are at distance 1 and the remaining $d_G(u)$ vertices are at distance at least 2 from u. Therefore

$$\begin{aligned} rs_{\overline{G}}(u) &\le (n-1-d_G(u)) + \frac{1}{2}d_G(u) \\ &= n - 1 - \frac{1}{2}d_G(u). \end{aligned}$$

Therefore

$$
\begin{aligned}
RS_1(\overline{G}) &= \sum_{uv \in E(\overline{G})} [rs_{\overline{G}}(u) + rs_{\overline{G}}(v)] \\
&\leq \sum_{uv \in E(\overline{G})} \left[(n-1) - \frac{1}{2}d_G(u) \right] + \left[(n-1) - \frac{1}{2}d_G(v) \right] \\
&= \sum_{uv \in E(\overline{G})} \left[2(n-1) - \frac{1}{2}(d_G(u) + d_G(v)) \right] \\
&= \sum_{uv \notin E(G)} 2(n-1) - \frac{1}{2} \sum_{uv \notin E(G)} [d_G(u) + d_G(v)] \\
&= 2(n-1)\left(\frac{n(n-1)}{2} - m \right) - \frac{1}{2}\overline{Z_1}(G) \\
&= n(n-1)^2 - 2m(n-1) - \frac{1}{2}\overline{Z_1}(G)
\end{aligned}
$$

and

$$
\begin{aligned}
RS_2(\overline{G}) &\leq \sum_{uv \in E(\overline{G})} [rs_{\overline{G}}(u) rs_{\overline{G}}(v)] \\
&= \sum_{uv \in E(\overline{G})} \left((n-1) - \frac{1}{2}d_G(u) \right) \left((n-1) - \frac{1}{2}d_G(v) \right) \\
&= \sum_{uv \in E(\overline{G})} \left((n-1)^2 - \frac{1}{2}(n-1)d_G(v) - \frac{(n-1)}{2}d_G(u) \right. \\
&\qquad \left. + \frac{1}{4}d_G(u)d_G(v) \right) \\
&= \sum_{uv \notin E(G)} (n-1)^2 - \frac{1}{2}(n-1) \sum_{uv \notin E(G)} (d_G(u) + d_G(v)) \\
&\qquad + \frac{1}{4} \sum_{uv \notin E(G)} (d_G(u)d_G(v)) \\
&= (n-1)^2 \left(\frac{n(n-1)}{2} - m \right) - \frac{1}{2}(n-1)\overline{Z_1}(G) + \frac{1}{4}\overline{Z_2}(G).
\end{aligned}
$$

□

2.6 Reciprocal Status Connectivity Indices and Co-indices of Some Reciprocal Status Regular Graphs

A graph G is said to be *reciprocal status regular* if reciprocal status $rs_G(u)$ of every vertex $u \in V(G)$ is the same. That is $rs_G(u) = k$, for each $u \in V(G)$.

A bijection $\propto$ on $V(G)$ is called automorphism of G if it preserves $E(G)$. In other words, $\propto$ is an automorphism if for each $u, v \in V(G)$, $e = uv \in E(G)$ if and only if $\propto (e) = \propto (u) \propto (v) \in E(G)$. Let

$$Aut(G) = \quad \{\propto \mid \propto \colon V(G) \to V(G)$$

$$\text{is a bijection, which preserves the adjacency}\}$$

It is known that $Aut(G)$ forms a group under the composition of mappings. A graph G is called vertex reciprocal status if for every two vertices u and v of G, there exists an automorphism $\propto$ of G such that $\propto (u) = \propto (v)$.

The following is straightforward from the definition of reciprocal status connectivity indices.

Lemma 2.6.1. *Let G be a connected k-reciprocal status regular graph with m edges. Then $RS_1(G) = 2mk$ and $RS_2(G) = mk^2$.*

43

Theorem 2.6.2. *Let G be a connected graph on n vertices with the automorphism group $Aut(G)$ and the vertex set $V(G)$.*

Let $V_1, V_2, \ldots, V_t$ be all orbits of the action $Aut(G)$ on $V(G)$. Suppose that for each $1 \leq i \leq t$, k_i are the reciprocal status of vertices in the orbit V_i, respectively. Then $HI(G) = \frac{1}{2} \sum_{i=1}^{t} |V_i| k_i$.

Specially if G is vertex reciprocal status regular (that is, $t = 1$), then $HI(G) = \frac{1}{2} nk$, where k is the reciprocal status of each vertex of G.

Analogous to Theorem 2.6.2 and as a consequence of Proposition 2.5.1, we have the following.

Theorem 2.6.3. *Let G be a connected graph on n vertices with the automorphism group $Aut(G)$ and the vertex set $V(G)$. Let $V_1, V_2, \ldots, V_t$ be all orbits of the action $Aut(G)$ on $V(G)$. Suppose that for each $1 \leq i \leq t$, d_i and k_i are the vertex degree and the reciprocal status of vertices in the orbit V_i, respectively. Then*

$RS_1(G) = \sum_{i=1}^{t} |V_i| d_i k_i$ *and* $\overline{RS_1}(G) = (n-1) \sum_{i=1}^{t} \left(|V_i| k_i (1 - \frac{d_i}{n-1}) \right).$

Specially if G is vertex reciprocal status regular (that is, $t = 1$), then $RS_1(G) = ndk$, $RS_2(G) = \frac{1}{2} ndk^2$,

$$\overline{RS_1}(G) = 2 \binom{n}{2} k - ndk,$$

and

$$\overline{RS_2}(G) = \left(\binom{n}{2} - \frac{nd}{2} \right) k^2,$$

where d and k are the degree and the reciprocal status of each vertex of G respectively.

The following is a direct consequence of Proposition 2.5.1, Lemma 2.6.1 and Theorem 2.6.2.

Lemma 2.6.4. *Let G be a connected k-reciprocal status regular graph with n vertices and m edges. Then $\overline{RS_1}(G) = 2\binom{n}{2}k - 2mk$ and $\overline{RS_2}(G) = \binom{n}{2}k^2 - mk^2$.*

The Kneser graph $KG_{p,k}$ is the graph whose vertices correspond to the k-element subsets of a set of p elements, and where two vertices are adjacent if and only if the two corresponding sets are disjoint. Clearly we must impose the restriction $p \geq 2k$. The Kneser graph $KG_{p,k}$ has $\binom{p}{k}$ vertices and it is regular of degree $\binom{p-k}{k}$. Therefore the number of edges of $KG_{p,k}$ is $\frac{1}{2}\binom{p}{k}\binom{p-k}{k}$ (see [39]). The Kneser graph $KG_{n,1}$ is the complete graph on n vertices. The Kneser graph $KG_{2p-1,p-1}$ is known as the odd graph O_p. The odd graph $O_3 = KG_{5,2}$ is isomorphic to the Peterson graph (see Fig. 2.6).

Figure 2.6 :The odd graph $O_3 = KG_{5,2}$ is isomorphic to the Peterson graph

Lemma 2.6.5. [39] *The Kneser graph $KG_{p,k}$ is vertex reciprocal status regular and for each k-subset A, $rs_{KG_{p,k}}(A) = \dfrac{2HI(KG_{p,k})}{\binom{p}{k}}$.*

Following Proposition follows from Lemmas 2.6.1 and 2.6.5.

Proposition 2.6.6. *For a Kneser graph $KG_{p,k}$ we have*

$$RS_1(KG_{p,k}) = 2HI(KG_{p,k})\binom{p-k}{k}$$

and

$$RS_2(KG_{p,k}) = \binom{p-k}{k}\left[\frac{2HI(KG_{p,k})^2}{\binom{p}{k}}\right].$$

Following Proposition follows from Proposition 2.5.1, Lemma 2.6.5 and Proposition 2.6.6.

Proposition 2.6.7. *For a Kneser graph $KG_{p,k}$ we have*

$$\overline{RS_1}(KG_{p,k}) = 2HI(KG_{p,k})\left[\binom{p}{k} - \binom{p-k}{k} - 1\right]$$

and

$$\overline{RS_2}(KG_{p,k}) = \left(\frac{4\left(HI(KG_{p,k})^2\right)}{\binom{p}{k}^2}Q\right),$$

where Q is the number of edges which does not belong to $E(KG_{p,k})$.

2.7 Harmonic Reciprocal Status Index

First we give bounds for the harmonic reciprocal status index.

Theorem 2.7.1. *Let G be a connected graph with n vertices and let $diam(G) = D$. Then*

$$\sum_{uv \in E(G)} \frac{2}{n - 1 + \frac{1}{2}\left[d_G(u) + d_G(v)\right]} \leq HRS(G)$$

$$\leq \sum_{uv \in E(G)} \frac{2}{\frac{2}{D}(n-1) + \left(1 - \frac{1}{D}\right)\left[d_G(u) + d_G(v)\right]}.$$

Equality on both sides holds if and only if $diam(G) \leq 2$.

Proof. <u>Upper bound:</u> For any vertex u of G, there are $d_G(u)$ vertices which are at distance 1 from u and the remaining $n-1-d_G(u)$ vertices are at distance at most D. Therefore for any vertex $u \in V(G)$,

$$rs_G(u) \geq d_G(u) + \frac{1}{D}(n-1-d_G(u)) = \frac{1}{D}(n-1) + d_G(u)\left(1 - \frac{1}{D}\right).$$

Therefore

$$\begin{aligned}
HRS(G) &= \sum_{uv \in E(G)} \frac{2}{rs_G(u) + rs_G(v)} \\
&\leq \sum_{uv \in E(G)} \frac{2}{\frac{2}{D}(n-1) + \left(1 - \frac{1}{D}\right)(d_G(u) + d_G(v))}.
\end{aligned}$$

<u>Lower bound:</u> For any vertex u of G, there are $d_G(u)$ vertices which are at distance 1 from u and the remaining $n-1-d_G(u)$ vertices are at distance at least 2. Therefore for any vertex $u \in V(G)$,

$$rs_G(u) \leq d_G(u) + \frac{1}{2}(n-1-d_G(u)) = \frac{1}{2}[d_G(u) + n - 1].$$

Therefore

$$\begin{aligned}
HRS(G) &= \sum_{uv \in E(G)} \frac{2}{rs_G(u) + rs_G(v)} \\
&\geq \sum_{uv \in E(G)} \frac{2}{(n-1) + \frac{1}{2}[d_G(u) + d_G(v)]}.
\end{aligned}$$

<u>For equality:</u> If the diameter of G is 1 or 2 then the equality holds.

Conversely, let

$$HRS(G) = \sum_{uv \in E(G)} \frac{2}{(n-1) + \frac{1}{2}[d_G(u) + d_G(v)]}.$$

Suppose, $diam(G) \geq 3$, then there exists at least one pair of vertices, say u_1 and u_2 such that $d_G(u_1, u_2) \geq 3$.

Therefore

$$rs_G(u_1) \leq d_G(u_1) + \frac{1}{3} + \frac{1}{2}(n - 2 - d_G(u_1)) = \frac{n}{2} - \frac{2}{3} + \frac{d_G(u_1)}{2}.$$

Similarly $rs_G(u_2) \leq \frac{n}{2} - \frac{2}{3} + \frac{d_G(u_2)}{2}$ and for all other vertices u of G, $rs_G(u) \leq \frac{n}{2} - \frac{1}{2} + \frac{d_G(u)}{2}$.

Partition the edge set of G into three sets E_1, E_2 and E_3, such that

$$E_1 = \left\{ u_1 v \mid rs_G(u_1) \leq \frac{n}{2} - \frac{2}{3} + \frac{d_G(u_1)}{2} \text{ and } rs_G(v) \leq \frac{n}{2} - \frac{1}{2} + \frac{d_G(v)}{2} \right\},$$

$$E_2 = \left\{ u_2 v \mid rs_G(u_2) \leq \frac{n}{2} - \frac{2}{3} + \frac{d_G(u_2)}{2} \text{ and } rs_G(v) \leq \frac{n}{2} - \frac{1}{2} + \frac{d_G(v)}{2} \right\}$$

and

$$E_3 = \left\{ uv \mid rs_G(u) \leq \frac{n}{2} - \frac{1}{2} + \frac{d_G(u)}{2} \text{ and } rs_G(v) \leq \frac{n}{2} - \frac{1}{2} + \frac{d_G(v)}{2} \right\}.$$

It is easy to check that $|E_1| = d_G(u_1)$, $|E_2| = d_G(u_2)$ and $|E_3| = m - d_G(u_1) - d_G(u_2)$. Therefore

$$
\begin{aligned}
HRS(G) &= \sum_{uv \in E(G)} \frac{2}{rs_G(u) + rs_G(v)} \\
&= \sum_{u_1 v \in E_1} \frac{2}{rs_G(u_1) + rs_G(v)}
\end{aligned}
$$

$$+ \sum_{u_2 v \in E_2} \frac{2}{rs_G(u_2) + rs_G(v)} + \sum_{uv \in E_3} \frac{2}{rs_G(u) + rs_G(v)}$$

$$\geq \sum_{u_1 v \in E_1} \frac{2}{\left[n - \frac{7}{6} + \frac{1}{2}(d_G(u_1) + d_G(v))\right]}$$

$$+ \sum_{u_2 v \in E_2} \frac{2}{\left[n - \frac{7}{6} + \frac{1}{2}(d_G(u_2) + d_G(v))\right]}$$

$$+ \sum_{uv \in E_3} \frac{2}{\left[n - 1 + \frac{1}{2}(d_G(u) + d_G(v))\right]}$$

$$> \sum_{uv \in E(G)} \frac{2}{n - 1 + \frac{1}{2}[d_G(u) + d_G(v)]},$$

which is a contradiction. Hence $diam(G) \leq 2$. $\qquad\square$

Corollary 2.7.2. *Let G be a connected graph with n vertices, m edges and $diam(G) = D$. Let δ and Δ be the minimum and maximum degree of the vertices of G respectively. Then*

$$\frac{2m}{n - 1 + \Delta} \leq HRS(G) \leq \frac{m}{\frac{n-1}{D} + \left(1 - \frac{1}{D}\right)\delta}.$$

Proof. For any vertex u of G, $\delta \leq d_G(u) \leq \Delta$. Therefore substituting $d_G(u) + d_G(v) \geq 2\delta$ in the upper bound and $d_G(u) + d_G(v) \leq 2\Delta$ in the lower bound of Theorem 2.7.1, we get the results. $\qquad\square$

Corollary 2.7.3. *Let G be a connected regular graph of degree r on n vertices and m edges and let $diam(G) = D$. Then*

$$\frac{2m}{n - 1 + r} \leq HRS(G) \leq \frac{m}{\frac{n-1}{D} + \left(1 - \frac{1}{D}\right)r}.$$

Equality on both side holds if and only if $diam(G) \leq 2$.

Proof. For any vertex u of G, $d_G(u) = r$. Therefore the results follows by the Theorem 2.7.1. $\qquad\square$

Now we compute the harmonic reciprocal status index of some specific graphs.

Proposition 2.7.4. *For a complete graph K_n on n vertices,*

$$HRS(K_n) = \frac{n}{2}.$$

Proof. For any vertex u of K_n, $rs_{K_n}(u) = n - 1$. Therefore by the definition of harmonic reciprocal status index, $HRS(K_n) = \frac{n}{2}$. $\qquad\square$

Proposition 2.7.5. *For a complete bipartite graph $K_{p,q}$,*

$$HRS(K_{p,q}) = \frac{4pq}{3(p+q) - 2}.$$

Proof. The vertex set $V(K_{p,q})$ can be partitioned into two independent sets V_1 and V_2 such that for every edge uv of $K_{p,q}$, the vertex $u \in V_1$ and $v \in V_2$. Therefore $d_{K_{p,q}}(u) = q$ and $d_{K_{p,q}}(v) = p$, where $|V_1| = p$ and $|V_2| = q$. The graph $K_{p,q}$ has $n = p+q$ vertices and $m = pq$ edges. Also $diam(K_{p,q}) \leq 2$. Therefore by the equality part of Theorem 2.7.1,

$$HRS(K_{p,q}) = \sum_{uv \in E(K_{p,q})} \frac{2}{p + q - 1 + \frac{1}{2}[p+q]} = \frac{4pq}{3(p+q) - 2}.$$

$\qquad\square$

Proposition 2.7.6. *For a path P_n on n vertices,*

$$HRS(P_n) = \left[\frac{4}{\frac{n}{n-1} + 2\sum_{i=1}^{n-2} \frac{1}{i}} \right] + \sum_{i=2}^{n-2} \left[\frac{2}{\frac{n}{i(n-i)} + 2\left[\sum_{j=1}^{i-1} \frac{1}{j} + \sum_{j=1}^{n-i-1} \frac{1}{j} \right]} \right].$$

Proof. Let $v_1, v_2, \ldots, v_n$ be the vertices of P_n, where v_i is adjacent to v_{i+1}, $i = 1, 2, \ldots, n-1$. Therefore for $i = 1, 2, \ldots, n$

$$rs_{P_n}(v_1) = \sum_{i=1}^{n-1} \frac{1}{i},$$

$$rs_{P_n}(v_i) = \sum_{j=1}^{i-1} \frac{1}{j} + \sum_{j=1}^{n-i} \frac{1}{j}, \quad \text{for } 2 \le i \le n-1$$

$$\text{and} \quad rs_{P_n}(v_n) = \sum_{i=1}^{n-1} \frac{1}{i}.$$

Therefore,

$$HRS(P_n) = \sum_{uv \in E(P_n)} \frac{2}{rs_{P_n}(u) + rs_{P_n}(v)}$$

$$= \left[\frac{2}{rs_{P_n}(v_1) + rs_{P_n}(v_2)} \right] + \sum_{i=2}^{n-2} \left[\frac{2}{rs_{P_n}(v_i) + rs_{P_n}(v_{i+1})} \right]$$

$$+ \left[\frac{2}{rs_{P_n}(v_{n-1}) + rs_{P_n}(v_n)} \right]$$

$$= \left[\frac{2}{\sum_{i=1}^{n-1} \frac{1}{i} + 1 + \sum_{j=1}^{n-2} \frac{1}{j}} \right]$$

$$+ \sum_{i=2}^{n-2} \left[\frac{2}{\sum_{j=1}^{i-1} \frac{1}{j} + \sum_{j=1}^{n-i} \frac{1}{j} + \sum_{j=1}^{i} \frac{1}{j} + \sum_{j=1}^{n-i-1} \frac{1}{j}} \right]$$

$$+ \left[\frac{2}{\sum_{j=1}^{n-2} \frac{1}{j} + 1 + \sum_{i=1}^{n-1} \frac{1}{i}} \right]$$

$$= \left[\frac{4}{\frac{n}{n-1} + 2\sum_{i=1}^{n-2} \frac{1}{i}} \right]$$

$$+ \sum_{i=2}^{n-2} \left[\frac{2}{\frac{n}{i(n-i)} + 2\left[\sum_{j=1}^{i-1} \frac{1}{j} + \sum_{j=1}^{n-i-1} \frac{1}{j} \right]} \right].$$

Proposition 2.7.7. *For a cycle C_n on $n \geq 3$ vertices,*

$$HRS(C_n) = \begin{cases} \dfrac{n}{\frac{2}{n}+2\sum_{i=1}^{(n-2)/2}\frac{1}{i}}, & \text{if } n \text{ is even} \\[4ex] \dfrac{n}{2\sum_{i=1}^{(n-1)/2}\frac{1}{i}}, & \text{if } n \text{ is odd.} \end{cases}$$

Proof. Case (i): If n is even number then for any vertex u of C_n,

$$rs_{C_n}(u) = 2\left[1 + \frac{1}{2} + \cdots + \frac{1}{\frac{n-2}{2}}\right] + \frac{1}{\frac{n}{2}} = \frac{2}{n} + 2\sum_{i=1}^{(n-2)/2}\frac{1}{i}.$$

Therefore,

$$\begin{aligned} HRS(C_n) &= \sum_{uv \in E(C_n)} \frac{2}{rs_{C_n}(u) + rs_{C_n}(v)} \\ &= \sum_{uv \in E(C_n)} \left[\frac{2}{\frac{2}{n}+2\sum_{i=1}^{(n-2)/2}\frac{1}{i}+\frac{2}{n}+2\sum_{i=1}^{(n-2)/2}\frac{1}{i}}\right] \\ &= \frac{n}{\frac{2}{n}+2\sum_{i=1}^{(n-2)/2}\frac{1}{i}}. \end{aligned}$$

Case (ii): If n is odd then for any vertex u of C_n,

$$rs_{C_n}(u) = 2\left[1 + \frac{1}{2} + \cdots + \frac{1}{\frac{n-1}{2}}\right] = 2\sum_{i=1}^{(n-1)/2}\frac{1}{i}.$$

Therefore

$$\begin{aligned} HRS(C_n) &= \sum_{uv \in E(C_n)} \frac{2}{rs_{C_n}(u) + rs_{C_n}(v)} \\ &= \sum_{uv \in E(C_n)} \frac{2}{2\sum_{i=1}^{(n-1)/2}\frac{1}{i}+2\sum_{i=1}^{(n-1)/2}\frac{1}{i}} \\ &= \frac{n}{2\sum_{i=1}^{(n-1)/2}\frac{1}{i}}. \end{aligned}$$

Proposition 2.7.8. *For a wheel* W_{n+1}, $n \geq 3$,

$$HRS(W_{n+1}) = \frac{2n(5n+9)}{3n^2 + 12n + 9}.$$

Proof. Partition the edge set $E(W_{n+1})$ into two sets E_1 and E_2, such that

$$E_1 = \{uv \mid d_{W_{n+1}}(u) = n \ \text{ and } \ d_{W_{n+1}}(v) = 3\}$$

and

$$E_2 = \{uv \mid d_{W_{n+1}}(u) = 3 \ \text{ and } \ d_{W_{n+1}}(v) = 3\}.$$

It is easy to check that $|E_1| = |E_2| = n$. Also $diam(W_{n+1}) = 2$. Therefore by the equality part of Theorem 2.7.1,

$$
\begin{aligned}
HRS(W_{n+1}) &= \sum_{uv \in E(W_{n+1})} \frac{2}{n + \frac{1}{2}\left[d_{W_{n+1}}(u) + d_{W_{n+1}}(v)\right]} \\
&= \sum_{uv \in E_1} \frac{2}{n + \frac{1}{2}\left[d_{W_{n+1}}(u) + d_{W_{n+1}}(v)\right]} \\
&\quad + \sum_{uv \in E_2} \frac{2}{n + \frac{1}{2}\left[d_{W_{n+1}}(u) + d_{W_{n+1}}(v)\right]} \\
&= \sum_{uv \in E_1} \frac{2}{n + \frac{1}{2}(n+3)} + \sum_{uv \in E_2} \frac{2}{n + \frac{1}{2}(3+3)} \\
&= \frac{2n}{n + \frac{1}{2}(n+3)} + \frac{2n}{n+3} \\
&= \frac{2n(5n+9)}{3n^2 + 12n + 9}.
\end{aligned}
$$

$\square$

Proposition 2.7.9. *For a friendship graph F_n, $n \geq 2$,*

$$HRS(F_n) = \frac{n(7n+5)}{3n^2 + 4n + 1}.$$

Proof. Partition the edge set $E(F_n)$ into two sets E_1 and E_2, such that $E_1 = \{uv \mid d_{F_n}(u) = 2n \text{ and } d_{F_n}(v) = 2\}$ and $E_2 = \{uv \mid d_{F_n}(u) = 2 \text{ and } d_{F_n}(v) = 2\}$. It is easy to check that $|E_1| = 2n$ ans $|E_2| = n$. Also $diam(F_n) = 2$. Therefore by the equality part of Theorem 2.7.1,

$$
\begin{aligned}
HRS(F_n) &= \sum_{uv \in E(F_n)} \frac{2}{2n + \frac{1}{2}[d_{F_n}(u) + d_{F_n}(v)]} \\
&= \sum_{uv \in E_1} \frac{2}{2n + \frac{1}{2}[d_{F_n}(u) + d_{F_n}(v)]} \\
&\quad + \sum_{uv \in E_2} \frac{2}{2n + \frac{1}{2}[d_{F_n}(u) + d_{F_n}(v)]} \\
&= \sum_{uv \in E_1} \frac{2}{2n + \frac{1}{2}[2n + 2]} + \sum_{uv \in E_2} \frac{2}{2n + \frac{1}{2}[2 + 2]} \\
&= \frac{4n}{3n + 1} + \frac{2n}{2n + 2} \\
&= \frac{n(7n+5)}{3n^2 + 4n + 1}.
\end{aligned}
$$

$\square$

2.8 Harmonic Reciprocal Status Co-index of Graphs

Theorem 2.8.1. *Let G be a connected graph on n vertices and let $diam(G) = D$. Then,*

$$\sum_{uv \notin E(G)} \frac{2}{n - 1 + \frac{1}{2}[d_G(u) + d_G(v)]} \leq \overline{HRS}(G)$$

$$\leq \sum_{uv \notin E(G)} \frac{2}{\frac{2}{D}(n - 1) + \left(1 - \frac{1}{D}\right)[d_G(u) + d_G(v)]}.$$

Equality on both sides holds if and only if $diam(G) \leq 2$.

Proof. Proof is analogous to that of Theorem 2.7.1. $\qquad\square$

Corollary 2.8.2. *Let G be a connected graph with n vertices, m edges and $diam(G) = D$. Let δ and Δ be the minimum and maximum degree of the vertices of G respectively. Then*

$$\frac{n(n - 1) - 2m}{n - 1 + \Delta} \leq \overline{HRS}(G) \leq \frac{n(n - 1) - 2m}{2\left[\frac{n-1}{D} + \left(1 - \frac{1}{D}\right)\delta\right]}.$$

Proof. For any vertex $u \in V(G)$, $\delta \leq d_G(u) \leq \Delta$. Therefore $2\delta \leq d_G(u) + d_G(v) \leq 2\Delta$. The graph G has $\frac{n(n-1)}{2} - m$ pair of non adjacent vertices. Substituting $d_G(u) + d_G(v) \geq 2\delta$ in the upper bound and $d_G(u) + d_G(v) \leq 2\Delta$ in the lower bound of Theorem 2.8.1 we get the results. $\qquad\square$

Corollary 2.8.3. *Let G be a connected r-regular graph on n vertices and let $diam(G) = D$. Then*

$$\frac{n(n - 1) - nr}{n - 1 + r} \leq \overline{HRS}(G) \leq \frac{n(n - 1) - nr}{2\left[\frac{n-1}{D} + \left(1 - \frac{1}{D}\right)r\right]}.$$

Equality on both sides holds if and only if $diam(G) \leq 2$.

Proof. Substituting $d_G(u) = r$ for all $u \in V(G)$ in Theorem 2.8.1, we get the results. $\qquad \square$

Proposition 2.8.4. *For a complete graph K_n, $\overline{HRS}(K_n) = 0$.*

Proposition 2.8.5. *For a complete bipartite graph $K_{p,q}$,*

$$\overline{HRS}(K_{p,q}) = \frac{p(p-1)}{2q+p-1} + \frac{q(q-1)}{2p+q-1}.$$

Proof. Let V_1 and V_2 be the partite sets of $V(K_{p,q})$, where $|V_1| = p$ and $|V_2| = q$ such that every edge of $K_{p,q}$ has one end in V_1 and other end in V_2. If $u \in V_1$ then $rs_{K_{p,q}}(u) = q + \frac{1}{2}(p-1)$ and if $u \in V_2$ then $rs_{K_{p,q}}(u) = p + \frac{1}{2}(q-1)$. Therefore for $u, v \in V_1$, $rs_{K_{p,q}}(u) + rs_{K_{p,q}}(v) = 2q + (p-1)$ and for $u, v \in V_2$, $rs_{K_{p,q}}(u) + rs_{K_{p,q}}(v) = 2p + (q-1)$. Therefore,

$$\begin{aligned}
\overline{HRS}(K_{p,q}) &= \sum_{uv \notin E(K_{p,q})} \frac{2}{rs_{K_{p,q}}(u) + rs_{K_{p,q}}(v)} \\
&= \sum_{\{u,v\} \subseteq V_1} \frac{2}{rs_{K_{p,q}}(u) + rs_{K_{p,q}}(v)} \\
&\quad + \sum_{\{u,v\} \subseteq V_2} \frac{2}{rs_{K_{p,q}}(u) + rs_{K_{p,q}}(v)} \\
&= \frac{p(p-1)}{2q+p-1} + \frac{q(q-1)}{2p+q-1}.
\end{aligned}$$

$\qquad \square$

Proposition 2.8.6. *For a cycle C_n on $n \geq 3$ vertices,*

$$
\overline{HRS}(C_n) = \begin{cases} \dfrac{n^2-3n}{\frac{4}{n}+4\sum_{i=1}^{(n-2)/2}\frac{1}{i}}, & \text{if } n \text{ is even} \\[3ex] \dfrac{n^3-3n}{4\sum_{i=1}^{(n-1)/2}\frac{1}{i}}, & \text{if } n \text{ is odd.} \end{cases}
$$

Proof. There are $\frac{n(n-1)}{2} - n$ pairs of non-adjacent vertices in C_n. As seen in Proposition 2.7.7, we have for a vertex u of C_n,

$$
rs_{C_n}(u) = \begin{cases} \dfrac{2}{n}+2\sum_{i=1}^{(n-2)/2}\frac{1}{i}, & \text{if } n \text{ is even} \\[3ex] 2\sum_{i=1}^{(n-1)/2}\frac{1}{i}, & \text{if } n \text{ is odd.} \end{cases}
$$

Therefore by the definition of harmonic reciprocal status co-index, we get the results. $\qquad \square$

Proposition 2.8.7. *For a wheel W_{n+1}, $n \geq 3$,*

$$
\overline{HRS}(W_{n+1}) = \frac{n(n-3)}{n+3}.
$$

Proof. The non adjacent pairs of vertices of the wheel W_{n+1} has degree 3 and there are $\frac{(n+1)n}{2} - 2n$ pairs of non adjacent vertices in W_{n+1}. Also $diam(W_{n+1}) = 2$. Therefore by the equality part of Theorem 2.8.1, we get the result. $\qquad \square$

Proposition 2.8.8. *For a friendship graph F_n, $n \geq 2$,*

$$
\overline{HRS}(F_n) = \frac{2n(n-1)}{n+1}.
$$

Proof. The non adjacent pairs of vertices of the windmill graph F_n has degree 2 and there are $\frac{2n(2n+1)}{2} - 3n$ such pairs in F_n. Also $diam(F_n) = $

2. Therefore by the equality part of Theorem 2.8.1, we get the result.

□

Chapter 3

Status Connectivity Index, Reciprocal Status Connectivity Index and Harmonic Reciprocal Status Index of Line Graphs

Results of this chapter are published in:

S. Y. Talwar and H. S. Ramane, "Reciprocal status and harmonic reciprocal status index of line graphs", Journal of Computer and Mathematical Sciences, 10(8) (2019), 1529-1538.(India) [ISSN: 0976 5727].

3.1 Introduction

Let $L(G)$ be the line graph of G. If $e = uv$ is an edge of G, then degree of e in $L(G)$ is $d_{L(G)}(e) = d_G(u) + d_G(v) - 2$.

Dobrynin and Kochetova [14] and Gutman [23] introduced a vertex-degree-weighted version of Wiener index called *degree distance* of G and is defined as:

$$DD(G) = \sum_{\{u,v\} \subseteq V(G)} [d_G(u) + d_G(v)] \, d_G(u,v) = \sum_{u \in V(G)} d_G(u)\sigma_G(u).$$
$$(3.1)$$

The Eq. (3.1) can be expressed as

$$DD(G) = \sum_{uv \in E(G)} [\sigma_G(u) + \sigma_G(v)], \qquad (3.2)$$

which is the first status connectivity index of G. Thus $S_1(G) = DD(G)$.

Bounds for the degree distance of a graph are obtained in [2, 8, 10, 35]. Extremal degree distance is studied in [6, 63, 65]. Degree distance of unicyclic and bicyclic graphs are reported in [17, 33, 64]. Degree distance of certain graphs are obtain in [1, 32, 59]. If T is a tree of order n, then [23]

$$S_1(G) = 4W(T) - n(n-1).$$

In [52] the first status connectivity index of line graph is obtained when $diam(L(G)) = 2$. In this chapter we obtain further results on the distance based indices of line graphs.

The forgotten topological index was studied at the earliest in [26] and is defined as,

$$F = F(G) = \sum_{v \in V(G)} d_G(v)^3 = \sum_{uv \in E(G)} \left[d_G(u)^2 + d_G(v)^2 \right].$$

This index was sighted off during the past and hence very less literature is available for the readers. One can go through [21].

The reciprocal status of a vertex $u \in V(G)$ and *First reciprocal status connectivity index* $RS_1(G)$ of a connected graph G are defined in Chapter 2 by Eqs. (2.1) and (2.3)

Harmonic reciprocal status index and co-index of a connected graph G are defined by Eqs. (2.7) and (2.8) in Chapter 2.

We need following results.

Theorem 3.1.1. *[25] Let G be a graph with n vertices and m edges. Then,*

$$Z_1(\overline{G}) = Z_1(G) + n(n-1)^2 - 4m(n-1).$$

Proposition 3.1.2. *[25] Let $L(G)$ be the line graph of the graph G. Then*

$$Z_1(L(G)) = F(G) - 4Z_1(G) + 2Z_2(G) + 4m,$$

where $Z_1(G)$, $Z_2(G)$ and $F(G)$ are the first Zagreb index, second Zagreb index and forgotton topological index of G respectively.

Theorem 3.1.3. *[50] For a connected graph G, $diam(L(G)) \leq 2$ if and only if none of the three graphs F_1, F_2 and F_3 of Fig. 3.1 are an induced subgraph of G.*

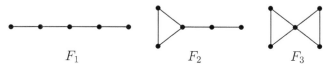

F_1 $\quad\quad$ F_2 $\quad\quad$ F_3

Figure 3.1 : Forbidden induced subgraphs.

3.2 First Status Connectivity Index of Line Graph of a Tree

Lemma 3.2.1. *Let $e = uv$ be an edge of a tree T and let T_1 and T_2 be the components of $T - e$, where $u \in V(T_1)$ and $v \in V(T_2)$. Then*

$$\sum_{f \in V(L(T_1))} d_{L(T)}(e, f) = \sum_{w \in V(T_1)} d_T(u, w)$$

and

$$\sum_{f \in V(L(T_2))} d_{L(T)}(e, f) = \sum_{w \in V(T_2)} d_T(v, w).$$

Proof. Every pair of vertices in a tree is joined by a unique path. Further an edge $e = uv$ is on every path whose one end is in $V(T_1)$ and other end is in $V(T_2)$. Therefore for $f = wx \in E(T_1)$, if $wfx \cdots uev$ is a path in T with end vertices w and v, then $d_{L(T)}(e, f) = d_T(u, w)$.

Hence

$$\sum_{f \in V(L(T_1))} d_{L(T)}(e, f) = \sum_{w \in V(T_1)} d_T(u, w).$$

Similarly we can show that

$$\sum_{f \in V(L(T_2))} d_{L(T)}(e, f) = \sum_{w \in V(T_2)} d_T(v, w).$$

□

Theorem 3.2.2. *Let T be a tree on n vertices and let $e = uv$ be an its edge. Then the status of a vertex e in $L(G)$ is*

$$\sigma_{L(T)}(e) = \frac{1}{2}\left[\sigma_T(u) + \sigma_T(v) - n\right]. \tag{3.3}$$

Proof. Let T_1 and T_2 be the components of $T - e$, where $u \in V(T_1)$ and $v \in V(T_2)$. Therefore $V(T_1) \cup V(T_2) = V(T)$ and $V(T_1) \cap V(T_2) = \phi$. Further $|V(T_1)| + |V(T_2)| = n$ and $|E(T_1)| + |E(T_2)| = n - 1$. Therefore

$$
\begin{aligned}
\sigma_{L(T)}(e) &= \sum_{f \in V(L(T))} d_{L(T)}(e, f) \\
&= \sum_{f \in V(L(T_1))} d_{L(T)}(e, f) + \sum_{f \in V(L(T_2))} d_{L(T)}(e, f) \\
&= \sum_{w \in V(T_1)} d_T(u, w) + \sum_{w \in V(T_2)} d_T(v, w). \ (By \ Lemma \ 3.2.1) \tag{3.4}
\end{aligned}
$$

Now

$$\begin{aligned}
\sigma_T(u) &= \sum_{w \in V(T)} d_T(u, w) \\
&= \sum_{w \in V(T_1)} d_T(u, w) + \sum_{w \in V(T_2)} d_T(u, w) \\
&= \sum_{w \in V(T_1)} d_T(u, w) + |V(T_2)| + \sum_{w \in V(T_2)} d_T(v, w). \quad (3.5)
\end{aligned}$$

Similarly

$$\sigma_T(v) = |V(T_1)| + \sum_{w \in V(T_1)} d_T(u, w) + \sum_{w \in V(T_2)} d_T(v, w). \quad (3.6)$$

Adding Eqs. (3.5) and (3.6) we get

$$\sigma_T(u) + \sigma_T(v) = 2 \left[\sum_{w \in V(T_1)} d_T(u, w) + \sum_{w \in V(T_2)} d_T(v, w) \right] + n \quad (3.7)$$

Equation (3.3) follows by Eqs. (3.4) and (3.7). $\qquad \square$

Lemma 3.2.3. *Let* $\mathbb{P}_k$ *be the set of all distinct paths of length* $k \geq 1$ *in a graph* G. *Then*

$$|\mathbb{P}_1| = m \quad and \quad |\mathbb{P}_2| = \frac{1}{2}[Z_1(G) - 2m].$$

Proof. The set $\mathbb{P}_1$ contains all edges of G. Hence $|\mathbb{P}_1| = m$.

If $d_G(u)$ is the degree of a vertex u, then there are $\binom{d_G(u)}{2}$ different paths of length 2 with u as a middle vertex. Hence

$$|\mathbb{P}_2| = \sum_{u \in V(G)} \binom{d_G(u)}{2}$$

$$= \frac{1}{2} \left[\sum_{u \in V(G)} d_G(u)^2 - \sum_{u \in V(G)} d_G(u) \right]$$

$$= \frac{1}{2} \left[Z_1(G) - 2m \right].$$

$\square$

Note that $uvw \in \mathbb{P}_2$ means uvw is a path of length 2 with v as its middle vertex.

Theorem 3.2.4. *Let T be a tree on n vertices and m edges. Then*

$$S_1(L(T)) = \frac{1}{2} \sum_{uvw \in \mathbb{P}_2} [\sigma_T(u) + 2\sigma_T(v) + \sigma_T(w)]$$

$$- \frac{n}{2} [Z_1(T) - 2m]. \tag{3.8}$$

Proof. Let $e = uv$ and $f = vw$ be the adjacent edges of T. Then by Eq. (3.2) and Theorem 3.2.2

$$S_1(L(T)) = \sum_{ef \in E(L(T))} [\sigma_{L(T)}(e) + \sigma_{L(T)}(f)]$$

$$= \sum_{uvw \in \mathbb{P}_2} \left[\frac{1}{2} [\sigma_T(u) + \sigma_T(v) - n] + \frac{1}{2} [\sigma_T(v) + \sigma_T(w) - n] \right]$$

$$= \frac{1}{2} \sum_{uvw \in \mathbb{P}_2} [\sigma_T(u) + 2\sigma_T(v) + \sigma_T(w)] - \sum_{uvw \in \mathbb{P}_2} n$$

$$= \frac{1}{2} \sum_{uvw \in \mathbb{P}_2} [\sigma_T(u) + 2\sigma_T(v) + \sigma_T(w)] - n \sum_{v \in V(T)} \binom{d_T(v)}{2}$$

$$= \frac{1}{2} \sum_{uvw \in \mathbb{P}_2} [\sigma_T(u) + 2\sigma_T(v) + \sigma_T(w)] - \frac{n}{2} [Z_1(T) - 2m].$$

$\square$

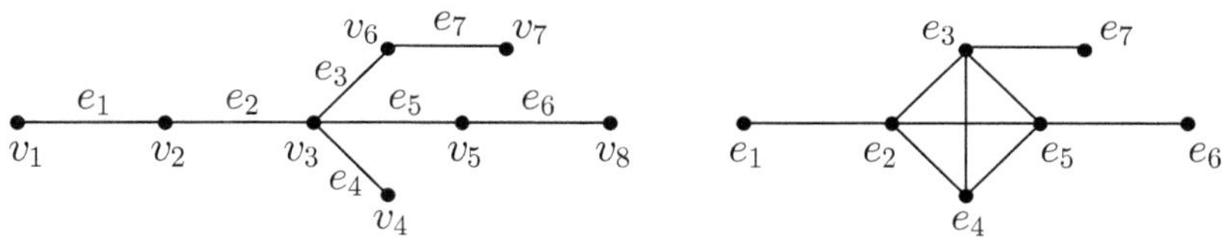

Figure 3.2 : Tree and its line graph.

Example: For a tree given in Fig. 3.2, $\sigma_T(v_1) = 20$, $\sigma_T(v_2) = 14$, $\sigma_T(v_3) = 10$, $\sigma_T(v_4) = 16$, $\sigma_T(v_5) = 14$, $\sigma_T(v_6) = 14$, $\sigma_T(v_7) = 20$, $\sigma_T(v_8) = 20$ and $Z_1(T) = 32$. The collection of all paths of length 2 is

$$\mathbb{P}_2 = \{v_1v_2v_3, v_2v_3v_4, v_2v_3v_5, v_2v_3v_6, v_3v_5v_8, v_3v_6v_7, v_4v_3v_5, v_4v_3v_6, v_5v_3v_6\}.$$

Therefore by Theorem 3.2.4,

$$\begin{aligned} S_1(L(T)) &= \frac{1}{2}[(20 + 2(14) + 10) + (14 + 2(10) + 16) \\ &\quad + (14 + 2(10) + 14) + (14 + 2(10) + 14) \\ &\quad + (10 + 2(14) + 20) + (10 + 2(14) + 20) \\ &\quad + (16 + 2(10) + 14) + (16 + 2(10) + 14) \\ &\quad + (14 + 2(10) + 14)] - \frac{8}{2}[32 - 2(7)] \\ &= 162. \end{aligned}$$

Corollary 3.2.5. *Let T be a tree with n vertices and m edges. Let k_u denotes the number of paths of length 2 with u as its end vertex and l_u denotes the number of paths of length 2 with u as its middle vertex. Then*

$$S_1(L(T)) = \frac{1}{2} \sum_{u \in V(T)} (k_u + 2l_u)\sigma_T(u) - \frac{n}{2}\left[Z_1(T) - 2m\right]. \qquad (3.9)$$

Proof. If k_u is the number of paths of length 2 with u as its end vertex and l_u is the number of paths of length 2 with u as its middle vertex then $\sigma_T(u)$ appeares $k_u + 2l_u$ times in $\sum_{uvw \in \mathbb{P}_2} [\sigma_T(u) + 2\sigma_T(v) + \sigma_T(w)]$. Therefore

$$\sum_{uvw \in \mathbb{P}_2} [\sigma_T(u) + 2\sigma_T(v) + \sigma_T(w)] = \sum_{u \in V(T)} (k_u + 2l_u)\sigma_T(u). \qquad (3.10)$$

Substituting Eq. (3.10) in Eq. (3.8) we get the Eq. (3.9).

$\square$

Theorem 3.2.6. *Let T be a tree on n vertices. Then*

$$S_1(L(T)) = \frac{1}{2} \sum_{u \in V(T)} d_T(u)^2 \sigma_T(u)$$

$$+ \frac{1}{2} \sum_{uv \in E(T)} [d_T(u)\sigma_T(v) + d_T(v)\sigma_T(u)]$$

$$- S_1(T) - \frac{n}{2} Z_1(T) + mn. \qquad (3.11)$$

Proof. Let $e = uv$ be an edge of T. Then by Eq. (3.1) and Theorem

3.2.2

$$S_1(L(T)) = \sum_{e \in V(L(T))} d_{L(T)}(e)\sigma_{L(T)}(e)$$

$$= \sum_{uv \in E(T)} (d_T(u) + d_T(v) - 2)\left(\frac{\sigma_T(u) + \sigma_T(v) - n}{2}\right)$$

$$= \frac{1}{2}\sum_{uv \in E(T)} [d_T(u)\sigma_T(u) + d_T(v)\sigma_T(v)]$$

$$+ \frac{1}{2}\sum_{uv \in E(T)} [d_T(u)\sigma_T(v) + d_T(v)\sigma_T(u)]$$

$$- \sum_{uv \in E(T)} [\sigma_T(u) + \sigma_T(v)] - \frac{n}{2}\sum_{uv \in E(T)} [d_T(u) + d_T(v)]$$

$$+ \sum_{uv \in E(T)} n$$

$$= \frac{1}{2}\sum_{u \in V(T)} d_T(u)^2 \sigma_T(u)$$

$$+ \frac{1}{2}\sum_{uv \in E(T)} [d_T(u)\sigma_T(v) + d_T(v)\sigma_T(u)]$$

$$- S_1(T) - \frac{n}{2}Z_1(T) + mn.$$

$\square$

3.3 Bounds for First Status Connectivity Index of Line Graphs

Theorem 3.3.1. *Let G be a connected graph with n vertices and m edges and let $D_L = diam(L(G))$. Let $e = uv$ be an edge of G. Then*

$$2m - d_G(u) - d_G(v) \le \sigma_{L(G)}(e) \le (m+1)D_L + (1 - D_L)(d_G(u) + d_G(v)) - 2.$$
$$(3.12)$$

Equality holds on both sides if and only if none of the three graphs F_1, F_2 and F_3 of Fig. 3.1 are an induced subgraph of G.

Proof. Let $e = uv$ be an edge of G. The vertex e is adjacent to $d_G(u) + d_G(v) - 2$ vertices in $L(G)$ and not adjacent to remaining $m - (d_G(u) + d_G(v) - 2) - 1 = m - d_G(u) - d_G(v) + 1$ vertices in $L(G)$. The distance between e and these non-adjacent vertices is atleast 2 and atmost D_L. Therefore

$$
\begin{aligned}
\sigma_{L(G)}(e) &\ge (d_G(u) + d_G(v) - 2) + 2(m - d_G(u) - d_G(v) + 1) \\
&= 2m - d_G(u) - d_G(v)
\end{aligned}
$$

and

$$
\begin{aligned}
\sigma_{L(G)}(e) &\le (d_G(u) + d_G(v) - 2) + D_L(m - d_G(u) - d_G(v) + 1) \\
&= (m+1)D_L + (1 - D_L)(d_G(u) + d_G(v)) - 2.
\end{aligned}
$$

For equality: if none of the three graphs F_1, F_2, F_3 of Fig. 3.1 are an induced subgraph of G then by Theorem 3.1.3, $D_L = diam(L(G)) \le 2$.

Hence

$$\sigma_{L(G)}(e) = (m+1)D_L + (1-D_L)(d_G(u) + d_G(v)) - 2$$
$$= 2m - d_G(u) - d_G(v).$$

Conversely, let $\sigma_{L(G)}(e) = 2m - d_G(u) - d_G(v)$.

Suppose $diam(L(G)) \geq 3$. Then there exists atleast one vertex f in $L(G)$ such that $d_{L(G)}(e, f) > 2$. Therefore

$$\sigma_{L(G)}(e) \geq (d_G(u) + d_G(v) - 2) + 2(m - d_G(u) - d_G(v)) + 3$$
$$= 2m - d_G(u) - d_G(v) + 1,$$

which is a contradiction. Hence $diam(L(G)) \leq 2$. Therefore none of the three graphs F_1, F_2 and F_3 of Fig. 3.1 are an induced subgraph of G. $\qquad\square$

Theorem 3.3.2. *Let G be a connected graph with n vertices and m edges and let $D_L = diam(L(G))$. Then*

$$S_1(L(G)) \geq 2m\left[Z_1(G) - 2m\right] - \sum_{uvw\in\mathbb{P}_2} \left[d_G(u) + 2d_G(v) + d_G(w)\right]$$
$$(3.13)$$

and

$$S_1(L(G)) \leq ((m+1)D_L - 2)\left[Z_1(G) - 2m\right]$$
$$+(1-D_L)\sum_{uvw\in\mathbb{P}_2} \left[d_G(u) + 2d_G(v) + d_G(w)\right]. \qquad (3.14)$$

Equality holds in both cases if and only if none of the three graphs F_1, F_2 and F_3 of Fig. 3.1 are an induced subgraph of G.

Proof. Let $e = uv$ and $f = vw$ be the adjacent edges of G. Then by Eq. (3.2) and Theorem 3.3.1

$$
\begin{aligned}
S_1(L(G)) &= \sum_{ef \in E(L(G))} \left[\sigma_{L(G)}(e) + \sigma_{L(G)}(f) \right] \\
&\geq \sum_{uvw \in \mathbb{P}_2} \left[(2m - d_G(u) - d_G(v)) + (2m - d_G(v) + d_G(w)) \right] \\
&= \sum_{uvw \in \mathbb{P}_2} 4m - \sum_{uvw \in \mathbb{P}_2} \left[d_G(u) + 2d_G(v) + d_G(w) \right] \\
&= 4m \sum_{v \in V(G)} \binom{d_G(v)}{2} - \sum_{uvw \in \mathbb{P}_2} \left[d_G(u) + 2d_G(v) + d_G(w) \right] \\
&= 2m \left[Z_1(G) - 2m \right] - \sum_{uvw \in \mathbb{P}_2} \left[d_G(u) + 2d_G(v) + d_G(w) \right]
\end{aligned}
$$

and

$$
\begin{aligned}
S_1(L(G)) &= \sum_{ef \in E(L(G))} \left[\sigma_{L(G)}(e) + \sigma_{L(G)}(f) \right] \\
&\leq \sum_{uvw \in \mathbb{P}_2} \left[((m+1)D_L + (1 - D_L)(d_G(u) + d_G(v)) - 2) \right. \\
&\qquad \left. + ((m+1)D_L + (1 - D_L)(d_G(v) + d_G(w)) - 2) \right] \\
&= \sum_{uvw \in \mathbb{P}_2} (2(m+1)D_L - 4) \\
&\quad + (1 - D_L) \sum_{uvw \in \mathbb{P}_2} \left[d_G(u) + 2d_G(v) + d_G(w) \right] \\
&= (2(m+1)D_L - 4) \sum_{v \in V(G)} \binom{d_G(v)}{2} \\
&\quad + (1 - D_L) \sum_{uvw \in \mathbb{P}_2} \left[d_G(u) + 2d_G(v) + d_G(w) \right] \\
&= ((m+1)D_L - 2) \left[Z_1(G) - 2m \right] \\
&\quad + (1 - D_L) \sum_{uvw \in \mathbb{P}_2} \left[d_G(u) + 2d_G(v) + d_G(w) \right].
\end{aligned}
$$

Equality follows from the equality part of Theorem 3.3.1. □

Corollary 3.3.3. *Let G be a connected graph with n vertices and m edges. Let k_u denotes the number of paths of length 2 with u as its end vertex and l_u denotes the number of paths of length 2 with u as its middle vertex. Let $D_L = diam(L(G))$. Then*

$$S_1(L(G)) \geq 2m\,[Z_1(G) - 2m] - \sum_{u \in V(G)} (k_u + 2l_u)d_G(u) \qquad (3.15)$$

and

$$S_1(L(G)) \leq ((m+1)D_L - 2)\,[Z_1(G) - 2m]$$

$$+(1 - D_L)\sum_{u \in V(G)} (k_u + 2l_u)d_G(u) \qquad (3.16)$$

Equality holds in both cases if and only if none of the three graphs F_1, F_2 and F_3 of Fig. 3.1 are an induced subgraph of G.

Proof. If k_u is the number of paths of length 2 with u as its end vertex and l_u is the number of paths of length 2 with u as its middle vertex then $d_G(u)$ appeares $k_u + 2l_u$ times in $\sum_{uvw \in \mathbb{P}_2} [d_G(u) + 2d_G(v) + d_G(w)]$. Therefore

$$\sum_{uvw \in \mathbb{P}_2} [d_G(u) + 2d_G(v) + d_G(w)] = \sum_{u \in V(G)} (k_u + 2l_u)d_G(u). \qquad (3.17)$$

Substituting Eq. (3.17) in Eqs. (3.13) and (3.14) we get the Eqs. (3.15) and (3.16).

Equality follows from the equality part of Theorem 3.3.1. □

If G is a connected r-regular graph then $Z_1(G) = nr^2$ and $m = nr/2$. Hence by Theorem 3.3.2, we have the following result.

Corollary 3.3.4. *Let G be a connected, r-regular graph on n vertices and let $D_L = diam(L(G))$. Then*

$$S_1(L(G)) \geq nr^2(r-1)(n-2) \tag{3.18}$$

and

$$S_1(L(G)) \leq nr(r-1)\left[\left(\frac{nr}{2}+1-2r\right)D_L + 2r - 2\right]. \tag{3.19}$$

Equality holds in both cases if and only if none of the three graphs F_1, F_2 and F_3 of Fig. 3.1 are an induced subgraph of G.

Let K_n be the complete graph on n vertices and $K_{p,q}$ be the complete bipartite graph on $p+q$ vertices. By Corollary 3.3.4, $S_1(L(K_n)) = n(n-1)^2(n-2)^2$ and $S_1(L(K_{p,p})) = 4p^3(p-1)^2$. A cocktail party graph H is a regular graph on $2k$ vertices and of degree $2k-2$. Hence $S_1(L(H)) = 16k(k-1)^3(2k-3)$.

Theorem 3.3.5. *Let G be a connected graph on n vertices and m edges. Let $D_L = diam(L(G))$. Then*

$$S_1(L(G)) \geq 2(m+1)Z_1(G) - 4m^2 - \sum_{uv \in E(G)} [d_G(u) + d_G(v)]^2 \tag{3.20}$$

and

$$S_1(L(G)) \leq [(m+3)D_L - 4]\, Z_1(G) + [4 - 2(m+1)D_L]\, m$$

$$+(1 - D_L) \sum_{uv \in E(G)} [d_G(u) + d_G(v)]^2. \tag{3.21}$$

Equality holds in both cases if and only if none of the three graphs F_1, F_2 and F_3 of Fig. 3.1 are an induced subgraph of G.

Proof. Let $e = uv$ be an edge of G. Then by Eq. (3.1) and by Theorem 3.3.1

$$
\begin{aligned}
S_1(L(G)) &= \sum_{e \in V(L(G))} d_{L(G)}(e) \sigma_{L(G)}(e) \\
&\geq \sum_{uv \in E(G)} (d_G(u) + d_G(v) - 2)(2m - d_G(u) - d_G(v)) \\
&= 2(m+1) \sum_{uv \in E(G)} [d_G(u) + d_G(v)] - 4 \sum_{uv \in E(G)} m \\
&\quad - \sum_{uv \in E(G)} [d_G(u) + d_G(v)]^2 \\
&= 2(m+1) Z_1(G) - 4m^2 - \sum_{uv \in E(G)} [d_G(u) + d_G(v)]^2
\end{aligned}
$$

and

$$
\begin{aligned}
S_1(L(G)) &= \sum_{e \in V(L(G))} d_{L(G)}(e) \sigma_{L(G)}(e) \\
&\leq \sum_{uv \in E(G)} (d_G(u) + d_G(v) - 2) \\
&\quad [(m+1)D_L + (1 - D_L)(d_G(u) + d_G(v)) - 2] \\
&= [(m+3)D_L - 4] \sum_{uv \in E(G)} [d_G(u) + d_G(v)] \\
&\quad + \sum_{uv \in E(G)} [4 - 2(m+1)D_L] \\
&\quad + (1 - D_L) \sum_{uv \in E(G)} [d_G(u) + d_G(v)]^2
\end{aligned}
$$

$$= [(m+3)D_L - 4] Z_1(G) + [4 - 2(m+1)D_L] m$$
$$+ (1 - D_L) \sum_{uv \in E(G)} [d_G(u) + d_G(v)]^2 .$$

Equality follows from the equality part of Theorem 3.3.1. $\qquad \square$

3.4 First Reciprocal Status Connectivity Index of Line Graphs

Theorem 3.4.1. *Let G be a connected graph with m edges such that none of the three graphs F_1, F_2 and F_3 of Fig. 3.1 are an induced subgraph of G and $L(G)$ is the line graph. Then*

$$RS_1(L(G)) = \frac{1}{2}(m-5)Z_1(G) + \frac{1}{2}F(G) + Z_2(G) - m(m-3).$$

Proof. From the definition of the line graphs [16], the number of vertices of $L(G)$ is $n_1 = m$ and the number of edges of $L(G)$ is,

$$m_1 = \frac{1}{2} \sum_{u \in V(G)} d_G(u)^2 - m$$
$$= \frac{1}{2} Z_1(G) - m.$$

By Theorem 3.1.3, none of the graphs F_1, F_2, F_3 of Fig. 3.1 are an induced subgraph of G, then $diam(L(G)) \le 2$. Therefore, by Theorem 2.2.1

$$RS_1(L(G)) = m_1(n_1 - 1) + \frac{1}{2} Z_1(L(G))$$
$$= (m-1) \left[\frac{1}{2} Z_1(G) - m \right] + \frac{1}{2} Z_1(L(G)).$$

Substituting $Z_1(L(G))$ from Proposition 3.1.2 in the above equation we get

$$RS_1(L(G)) = \frac{1}{2}(m-5)Z_1(G) + \frac{1}{2}F(G) + Z_2(G) - m(m-3).$$

$\square$

The following Corollary follows from the Theorem 3.4.1.

Corollary 3.4.2. *Let G be a connected, regular graph of degree r on n vertices and m edges. Let none of the graphs F_1, F_2, F_3 of Fig. 3.1 are an induced subgraph of G. Then*

$$RS_1(L(G)) = mr(m-5) + 2mr^2 - m(m-3).$$

Proof. For regular graph G of degree r with m edges, $Z_1(G) = 2mr$, $Z_2(G) = mr^2$ and $F(G) = 2mr^2$. Substituting these in Theorem 3.4.1 we get

$$RS_1(L(G)) = mr(m-5) + 2mr^2 - m(m-3).$$

$\square$

Proposition 3.4.3. *The first reciprocal status connectivity index of line graph of complete bipartite graph $K_{p,q}$, is*

$$RS_1(L(K_{p,q})) = \frac{pq}{2}\left[(p+q)(pq-5) + p^2 + q^2 + 6\right].$$

Proof. The graph $K_{p,q}$ has $n = p + q$ vertices and $m = pq$ edges. Also $diam(K_{p,q}) \leq 2$. The vertex set $V(K_{p,q})$ can be partitioned into two sets V_1 and V_2 such that for every edge uv of $K_{p,q}$, the vertex $u \in V_1$ and $v \in V_2$, where $|V_1| = p$ and $|V_2| = q$. Therefore $d_{K_{p,q}}(u) = q$ and $d_{K_{p,q}}(v) = p$ and hence $Z_1(K_{p,q}) = pq(p+q)$, $Z_2(K_{p,q}) = (pq)^2$ and $F(K_{p,q}) = pq(p^2 + q^2)$. From Theorem 3.4.1 we get,

$$
\begin{aligned}
RS_1(L(K_{p,q})) &= \frac{1}{2}(m-5)Z_1(K_{p,q}) + \frac{1}{2}F(K_{p,q}) + Z_2(K_{p,q}) - m(m-3) \\
&= \frac{1}{2}(m-5)pq(p+q) + \frac{1}{2}pq(p^2+q^2) + (pq)^2 - m(m-3) \\
&= \frac{pq}{2}\left[(p+q)(pq-5) + p^2 + q^2 + 6\right].
\end{aligned}
$$

$\square$

Theorem 3.4.4. *If G is any connected graph with n vertices, m edges, maximum degree $\Delta(G)$ and $diam(L(G)) = D_L$, then*

$$
\begin{aligned}
RS_1(L(G)) \geq{}& \frac{(Z_1(G) - 2m)(m-1)}{m - 2\Delta(G) + 3} \\
&+ \frac{(D_L - 1)(F(G) - 4Z_1(G) + 2Z_2(G) + 4m)}{m - 2\Delta(G) + 3}
\end{aligned}
$$

and

$$
\begin{aligned}
RS_1(L(G)) \leq{}& (m-1)\left(\frac{1}{2}Z_1(G) - m\right) \\
&+ \frac{1}{2}(F(G) - 4Z_1(G) + 2Z_2(G) + 4m).
\end{aligned}
$$

Proof. By Proposition 3.1.2 we have

$$
Z_1(L(G)) = F(G) - 4Z_1(G) + 2Z_2(G) + 4m.
$$

We know from Theorem 2.3.1 that

$$
\frac{2m}{D}(n-1) + \left(1 - \frac{1}{D}\right)Z_1(G) \leq RS_1(G). \tag{3.22}
$$

Now for any graph G, the number of vertices of its line graph, $L(G)$ is $n_1 = m$ and the number of edges is

$$m_1 = \frac{1}{2} \sum_{u \in V(G)} d_G(u)^2 - m = \frac{1}{2} Z_1(G) - m,$$

where $d_G(u)$ is the degree of the vertex u in G. Therefore, from Eq. (3.22) we have

$$RS_1(L(G)) \geq \frac{2m_1(m-1)}{D_L} + \left(1 - \frac{1}{D_L}\right) Z_1(L(G))$$

$$RS_1(L(G)) \geq 2\left[\frac{1}{2}Z_1(G) - m\right]\frac{m-1}{D_L} \tag{3.23}$$
$$+ \left(1 - \frac{1}{D_L}\right) Z_1(L(G))$$
$$= \frac{(Z_1(G) - 2m)(m-1)}{D_L} + \frac{(D_L - 1)Z_1(L(G))}{Z_1(L(G))},$$

where, D_L is the diameter of $L(G)$. For any graph G of order n, $diam(G) \leq n - \Delta(G) + 1$ and hence

$$\frac{1}{D_L} \geq \frac{1}{m - \Delta(L(G)) + 1}. \tag{3.24}$$

Also for any graph G,

$$\frac{1}{\Delta(L(G))} \leq \frac{1}{2(\Delta(G) - 1)}. \tag{3.25}$$

Substituting Eqs. 3.24 and 3.25 in Eq. 3.23 we get the following,

$$RS_1(L(G)) \geq \frac{(Z_1(G) - 2m)(m-1)}{m - 2\Delta(G) + 3}$$
$$+ \frac{(D_L - 1)(F(G) - 4Z_1(G) + 2Z_2(G) + 4m)}{m - 2\Delta(G) + 3}.$$

We know from Theorem 2.3.1 that

$$RS_1(G) \leq m(n-1) + \frac{1}{2}Z_1(G). \qquad (3.26)$$

For any graph G, the number of vertices of its line graph, $L(G)$ is $n_1 = m$ and the number of edges is

$$m_1 = \frac{1}{2}\sum_{u \in V(G)} d_G(u)^2 - m = \frac{1}{2}Z_1(G) - m.$$

Therefore, from Eq. (3.26) we have

$$\begin{aligned} RS_1(L(G)) &\leq m_1(m-1) + \frac{1}{2}Z_1(L(G)) \\ &= (m-1)\left(\frac{1}{2}Z_1(G) - m\right) \\ &\quad + \frac{1}{2}(F(G) - 4Z_1(G) + 2Z_2(G) + 4m). \end{aligned}$$

$\square$

Theorem 3.4.5. *Let G be a graph with n vertices and m edges. Let $diam(L(G)) > 3$. Then*

$$RS_1(\overline{L(G)}) = \frac{Z_1(G)}{2}(m+3) + \frac{1}{2}F(G) + Z_2(G) + m\left[\frac{1}{2}m^2 - 2m + \frac{7}{2}\right].$$

Proof. From Theorem 2.3.1 if $diam(G) \leq 2$, then

$$RS_1(G) = m(n-1) + \frac{1}{2}Z_1(G).$$

Since there exists a fact that for any graph G, if $diam(G) > 3$, then $diam(\overline{G}) \leq 2$. Therefore, if n_1, m_1 are the number of vertices and edges of $\overline{L(G)}$ respectively, then

$$
\begin{aligned}
RS_1(\overline{L(G)}) &= m_1(n_1 - 1) + \frac{1}{2}Z_1(\overline{L(G)}) \\
&= \left[\frac{1}{2}Z_1(G) - m\right](m - 1) + \frac{1}{2}Z_1(\overline{L(G)}).
\end{aligned}
$$

$$
\begin{aligned}
&= \left[\frac{1}{2}Z_1(G) - m\right](m - 1) + \frac{1}{2}\left[Z_1(L(G)) + m(m - 1)^2\right. \\
&\quad \left. - 4m_1(m - 1)\right] \quad (By\ Theorem\ 3.1.1) \\
&= \left[\frac{1}{2}Z_1(G) - m\right](m - 1) + \frac{1}{2}\left[Z_1(L(G)) + m(m - 1)^2\right. \\
&\quad \left. - 4\left(\frac{1}{2}Z_1(G) - m\right)(m - 1)\right] \\
&= \left[\frac{1}{2}Z_1(G) - m\right](m - 1) \\
&\quad + \frac{1}{2}\left[F(G) - 4Z_1(G) + 2Z_2(G) + 4m + m(m - 1)^2\right. \\
&\quad \left. - 4\left(\frac{1}{2}Z_1(G) - m\right)(m - 1)\right] \\
&= \frac{Z_1(G)}{2}(m + 3) + \frac{1}{2}F(G) + Z_2(G) + m\left[\frac{1}{2}m^2 - 2m + \frac{7}{2}\right].
\end{aligned}
$$

$\square$

3.5 Harmonic Reciprocal Status Index and Co-index of Line Graphs

Theorem 3.5.1. *Let G be a connected r-regular graph on n vertices and m edges and none of the three graphs F_1, F_2 and F_3 of Fig. 3.1* *are an induced subgraph of G then*

$$HRS(L(G)) = \frac{nr^2 - 2m}{m + 2r - 3}$$

and

$$\overline{HRS}(L(G)) = \frac{m(m - 2r + 1)}{(m + 2r - 3)}.$$

Proof. From the definition of the line graphs [16], the number of vertices of $L(G)$ is $n_1 = m$ and the number of edges is $m_1 = -m + \frac{1}{2}\sum_{u \in V(G)} d_G(u)^2$. Thus if G is r-regular, then $m_1 = -m + \frac{1}{2}nr^2$.

From Corollary 2.7.3

$$HRS(G) = \frac{2m}{(n-1) + r}. \tag{3.27}$$

Since G has no F_i, $i = 1, 2, 3$ as its induced subgraph, $diam(L(G)) \leq 2$. If G is r regular then $L(G)$ is $r_1 = 2r - 2$ regular graph. Therefore by Eq. (3.27),

$$\begin{aligned} HSR(L(G)) &= \frac{2m_1}{(n_1 - 1) + r_1} \\ &= \frac{2(-2m + nr^2)/2}{(m - 1) + 2r - 2} \\ &= \frac{-2m + nr^2}{m + 2r - 3}. \end{aligned}$$

From Corollary 2.8.3,

$$\overline{HRS}(G) = \frac{n(n-1) - nr}{(n-r) + r}.$$

Therefore

$$\overline{HRS}(L(G)) = \frac{n_1(n_1 - 1) - n_1 r_1}{(n_1 - 1) + r_1}$$
$$= \frac{m(m - 1) - m(2r - 2)}{(m - 1) + 2r - 2}$$
$$= \frac{m(m - 2r + 1)}{m + 2r - 3}.$$

$\square$

Theorem 3.5.2. *Let G be any connected graph with n vertices, m edges, δ and Δ be the minimum and maximum degrees of G respectively. Then*

$$HRS(L(G)) \geq \frac{-2m + Z_1(G)}{2\Delta - 3 + m}$$

and

$$\overline{HRS}(L(G)) \geq \frac{m^2 + m - Z_1(G)}{2\Delta - 3 + m}.$$

Proof. From the definition of the line graphs [16], the number of vertices of $L(G)$ is $n_1 = m$. Consider an edge $e = uv \in E(G)$ which is adjacent to $d_G(u) + d_G(v) - 2 = d_G(e)$ edges at u and v taken together in $L(G)$. Hence the edge e is not adjacent to remaining $m - 1 - d_G(e)$ edges of G. In $L(G)$ the distance between e and the remaining $m - 1 - d_G(e)$ vertices is at least 2. Hence for any graph G, $rs_{L(G)}(e)$ is

$$rs_{L(G)}(e) \leq [d_G(u) + d_G(v) - 2] + \frac{1}{2}(m - 1 - [d_G(u) + d_G(v) - 2]).$$

Since Δ is the maximum degree, $d_G(u) + d_G(v) \leq 2\Delta$. Therefore

$$rs_{L(G)}(e) \leq \frac{1}{2}(2\Delta - 3 + m).$$

Now suppose e and f are any two vertices of $L(G)$, then

$$rs_{L(G)}(e) + rs_{L(G)}(f) \leq (2\Delta - 3 + m).$$

Now, from Eq. (2.7) we have,

$$
\begin{aligned}
HRS(L(G)) &= \sum_{ef \in E(L(G))} \frac{2}{rs_{L(G)}(e) + rs_{L(G)}(f)} \\
&\geq \sum_{ef \in E(L(G))} \frac{2}{2\Delta - 3 + m} \\
&= \left(-m + \frac{1}{2} \sum_{i=1}^{n} d_i^2\right) \left(\frac{2}{2\Delta - 3 + m}\right) \\
&= \frac{-2m + Z_1(G)}{2\Delta - 3 + m}.
\end{aligned}
$$

Now, from Eq. (2.8) we have,

$$
\begin{aligned}
\overline{HRS}(L(G)) &= \sum_{ef \notin E(L(G))} \frac{2}{rs_{L(G)}(e) + rs_{L(G)}(f)} \\
&\geq \sum_{ef \notin E(L(G))} \frac{2}{2\Delta - 3 + m} \\
&\geq \left(\frac{n_1(n_1 - 1)}{2} - m_1\right) \left(\frac{2}{2\Delta - 3 + m}\right) \\
&= \left(\frac{m(m - 1)}{2} + m - \frac{1}{2} \sum_{i=1}^{n} d_i^2\right) \left(\frac{2}{2\Delta - 3 + m}\right) \\
&= \frac{m^2 + m - Z_1(G)}{2\Delta - 3 + m}.
\end{aligned}
$$

$\square$

Chapter 4

Status Like Topological Indices of Graphs and its Regression Analysis with Some Molecular Properties

4.1 Introduction

In this chapter we study the status and reciprocal status based topological indices of a graph and use it to correlate with the chemical properties of certain molecules.

The molecular graph is a simple graph, representing the carbon-atom skeleton of an organic molecule. Thus, the vertices of a molecular graph represent the carbon atoms, and its edges the carbon-carbon bonds.

In this Chapter, the *sum-connectivity index* of a graph G, denoted by $SC(G)$, is defined as in [74], *atom-bond connectivity index* of a graph G, proposed by Estrada et al. [18], the *augmented Zagreb index* of a graph G, proposed by Furtula et al. [20], the *arithmetic-geometric index* of a graph G, proposed by Shigehalli and Kanabur [60], and the *geometric-arithmetic index* was invented by Vukicević and Furtula [67] are used as frames of reference.

The *status* of a vertex $u \in V(G)$, denoted by $\sigma(u)$, is defined in [28, 51],

The *status sum-connectivity index* of a graph G [58], is defined as

$$T_{SC}(G) = \sum_{uv \in E(G)} \frac{1}{\sqrt{\sigma_G(u) + \sigma_G(v)}}.$$

The *status geometric-arithmetic index* of a graph G, denoted by

$T_{GA}(G)$, is defined as [42]

$$T_{GA}(G) = \sum_{uv \in E(G)} \frac{2\sqrt{\sigma_G(u)\sigma_G(v)}}{\sigma_G(u) + \sigma_G(v)}.$$

Now we define the following status based topological indices.

The *status arithmetic-geometric index* of a graph G, denoted by $T_{AG}(G)$, is defined as

$$T_{AG}(G) = \sum_{uv \in E(G)} \frac{\sigma_G(u) + \sigma_G(v)}{2\sqrt{\sigma_G(u)\sigma_G(v)}}. \tag{4.1}$$

The *status atom-bond connectivity index* of a graph G, denoted by $T_{ABC}(G)$, is defined as

$$T_{ABC}(G) = \sum_{uv \in E(G)} \sqrt{\frac{\sigma_G(u) + \sigma_G(v) - 2}{\sigma_G(u)\sigma_G(v)}}. \tag{4.2}$$

The *status augmented Zagreb index* of a graph G, denoted by $T_{AZ}(G)$, is defined as

$$T_{AZ}(G) = \sum_{uv \in E(G)} \left[\frac{\sigma_G(u)\sigma_G(v)}{\sigma_G(u) + \sigma_G(v) - 2} \right]^3. \tag{4.3}$$

The *first reciprocal status connectivity index* $RS_1(G)$ and *second reciprocal status connectivity index* $RS_2(G)$ are defined in Chapter 2 by Eqs. (2.3) and (2.4). Now we define the following topological indices.

The *reciprocal status arithmatic-geometric index* of a graph G is denoted by $RT_{AG}(G)$ and it is defined as

$$RT_{AG}(G) = \sum_{uv \in E(G)} \frac{rs(u) + rs(v)}{2\sqrt{rs_G(u)rs_G(v)}}. \qquad (4.4)$$

The *reciprocal status geometric-arithmetic index* of a graph G is denoted by $RT_{GA}(G)$ and it is defined as

$$RT_{GA}(G) = \sum_{uv \in E(G)} \frac{2\sqrt{rs_G(u)rs_G(v)}}{rs_G(u) + rs_G(v)}. \qquad (4.5)$$

The *reciprocal status sum-connectivity index* of a graph G is denoted by $RT_{SC}(G)$ and it is defined as

$$RT_{SC}(G) = \sum_{uv \in E(G)} \frac{1}{\sqrt{rs_G(u) + rs_G(v)}}. \qquad (4.6)$$

The *reciprocal status atom-bond connectivity index* of a graph G is denoted by $RT_{ABC}(G)$ and it is defined as

$$RT_{ABC}(G) = \sum_{uv \in E(G)} \sqrt{\frac{rs_G(u) + rs_G(v) - 2}{rs_G(u)rs_G(v)}}. \qquad (4.7)$$

The *reciprocal status augmented Zagreb index* of a graph G is denoted by $RT_{AZ}(G)$ and it is defined as

$$RT_{AZ}(G) = \sum_{uv \in E(G)} \left[\frac{rs_G(u)rs_G(v)}{rs_G(u) + rs_G(v) - 2} \right]^3. \qquad (4.8)$$

Example : For a graph given in Fig. 4.1, $T_{SC}(G) \approx 1.435$, $T_{GA}(G) \approx 4.958$, $T_{AG}(G) \approx 5.042$, $T_{ABC}(G) \approx 2.643$, $T_{AZ}(G) \approx$

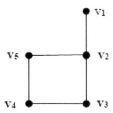

Figure 4.1 : Graph G.

233.472, $RT_{SC}(G) \approx 2.026$, $RT_{GA}(G) \approx 4.972$, $RT_{AG}(G) \approx 5.027$, $RT_{ABC}(G) \approx 3.337$, and $RT_{AZ}(G) \approx 56.810$.

Other reciprocal status based topological indices can be found in Chapter 1. In the next section we obtain bounds for the status based topological indices and reciprocal status based topological indices.

4.2 Bounds for Status and Reciprocal Status Based Topological Indices

Theorem 4.2.1. *Let G be a connected graph with n vertices and let $D = diam(G)$. Then,*

$$\sum_{uv \in E(G)} \frac{1}{\sqrt{2D(n-1) - (D-1)(d_G(u) + d_G(v))}} \leq T_{SC}(G)$$

$$\leq \sum_{uv \in E(G)} \frac{1}{\sqrt{4n - 4 - (d_G(u) + d_G(v))}}. \qquad (4.9)$$

Equality holds on both sides if and only if $diam(G) \leq 2$.

Proof. For any vertex u of G there are $d_G(u)$ vertices which are at distance 1 from u. Further, the distance between u and remaining $n - 1 - d_G(u)$ vertices is at least 2 and at most D. Therefore,

$$\sigma_G(u) \leq d_G(u) + D(n - 1 - d_G(u)) = D(n - 1) - (D - 1)d_G(u)$$

and

$$\sigma_G(u) \geq d_G(u) + 2(n - 1 - d_G(u)) = 2n - 2 - d_G(u),$$

with equality in both cases if and only if $D = 2$. Therefore,

$$4n - 4 - (d_G(u) + d_G(v)) \leq \sigma_G(u) + \sigma_G(v)$$
$$\leq 2D(n - 1) - (D - 1)(d_G(u) + d_G(v)).$$

Hence

$$T_{SC}(G) = \sum_{uv \in E(G)} \frac{1}{\sqrt{\sigma_G(u) + \sigma_G(v)}}$$
$$\geq \sum_{uv \in E(G)} \frac{1}{\sqrt{2D(n - 1) - (D - 1)(d_G(u) + d_G(v))}}$$

and

$$T_{SC}(G) = \sum_{uv \in E(G)} \frac{1}{\sqrt{\sigma_G(u) + \sigma_G(v)}}$$
$$\leq \sum_{uv \in E(G)} \frac{1}{\sqrt{4n - 4 - (d_G(u) + d_G(v))}}$$

Equality holds in both cases if and only if $D = 2$. $\square$

Theorems 4.2.2 to 4.2.4 can be proved analogous to the Theorem 4.2.1.

Theorem 4.2.2. *Let G be a connected graph with n vertices and let $D = diam(G)$. Then,*

$$\sum_{uv \in E(G)} \sqrt{\frac{2D(n-1) - (D-1)(d_G(u) + d_G(v)) - 2}{(D(n-1))^2 - D(n-1)(D-1)(d_G(u) + d_G(v)) - (D-1)^2 d_G(u) d_G(v)}} \leq T_{ABC}(G)$$

$$\leq \sum_{uv \in E(G)} \sqrt{\frac{4n - 6 - (d_G(u) + d_G(v))}{4n^2 - 8n - 4 + (2 - 2n)(d_G(u) + d_G(v)) + d_G(u) d_G(v)}}. \qquad (4.10)$$

Theorem 4.2.3. *Let G be a connected graph with n vertices and let $D = diam(G)$. Then,*

$$\sum_{uv \in E(G)} \frac{2D(n-1) - (D-1)(d_G(u) + d_G(v))}{\sqrt{2\left[(D(n-1))^2 - D(n-1)(D-1)(d_G(u) + d_G(v)) + (D-1)^2 d_G(u) d_G(v)\right]}} \leq T_{AG}(G)$$

$$\leq \sum_{uv \in E(G)} \frac{4n - 4 - (d_G(u) + d_G(v))}{2\sqrt{(2n - 2 - d_G(u))(2n - 2 - d_G(v))}}. \qquad (4.11)$$

Theorem 4.2.4. *Let G be a connected graph with n vertices and let $D = diam(G)$. Then,*

$$\sum_{uv \in E(G)} \left[\frac{(D(n-1))^2 - D(n-1)(D-1)(d_G(u) + d_G(v)) - (D-1)^2 d_G(u)d_G(v)}{2D(n-1) + (D-1)(d_G(u) + d_G(v)) - 2} \right]^3 \leq T_{AZ}(G)$$

$$\leq \sum_{uv \in E(G)} \left[\frac{4n^2 - 8n - 4 + (2 - 2n)(d_G(u) + d_G(v)) + d_G(u)d_G(v)}{4n - 6 - (d_G(u) + d_G(v))} \right] \quad (4.12)$$

Theorem 4.2.5. *Let G be a connected graph with n vertices and let $diam(G) = D$. Then,*

$$\sum_{uv \in E(G)} \frac{1}{\sqrt{(n-1) + \frac{1}{2}(d_G(u) + d_G(v))}} \leq RT_{SC}(G)$$

$$\leq \sum_{uv \in E(G)} \frac{1}{\sqrt{\frac{2(n-1)}{D} + \left(1 - \frac{1}{D}\right)(d_G(u) + d_G(v))}}. \quad (4.13)$$

Proof. **Lower bound :** For any vertex u of G there are $d_G(u)$ vertices which are at distance 1 from u and remaining $n - 1 - d_G(u)$ vertices are at distance at least 2. Therefore

$$rs_G(u) \leq \frac{1}{2}(n - 1 + d_G(u))$$

and

$$rs_G(u) + rs_G(v) \leq (n - 1) + \frac{1}{2}(d_G(u) + d_G(v)).$$

Hence

$$\begin{aligned} RT_{SC}(G) &= \sum_{uv \in E(G)} \frac{1}{\sqrt{rs_G(u) + rs_G(v)}} \\ &\geq \sum_{uv \in E(G)} \frac{1}{\sqrt{(n - 1) + \frac{1}{2}(d_G(u) + d_G(v))}}. \end{aligned}$$

Upper bound: For any vertex u of G there are $d_G(u)$ vertices which are at distance 1 from u and remaining $n - 1 - d_G(u)$ vertices are at distance at most D. Therefore,

$$rs_G(u) \geq \frac{1}{D}(n - 1) + (1 - \frac{1}{D})d_G(u)$$

and

$$rs_G(u) + rs_G(v) \geq \frac{2(n - 1)}{D} + \left(1 - \frac{1}{D}\right)(d_G(u) + d_G(v)).$$

Hence

$$RT_{SC}(G) \leq \sum_{uv \in E(G)} \frac{1}{\sqrt{\frac{2(n-1)}{D} + \left(1 - \frac{1}{D}\right)(d_G(u) + d_G(v))}}.$$

$\square$

In the following Theorem we obtained bounds for $RT_{ABC}(G)$, $RT_{AZ}(G)$, $RT_{AG}(G)$ and $RT_{GA}(G)$

Theorem 4.2.6. *Let G be a connected graph with n vertices and let $diam(G) = D$. Then,*

$$(i) \sum_{uv \in E(G)} \sqrt{\frac{2\left(2(n-1) + d_G(u) + d_G(v) - 4\right)}{(n-1)^2 + (n-1)(d_G(u) + d_G(v)) + d_G(u)d_G(v)}}$$

$$\leq RT_{ABC}(G) \leq$$

$$\sum_{uv \in E(G)} \sqrt{\frac{\frac{2(n-1)}{D} + \left(1 - \frac{1}{D}\right)(d_G(u) + d_G(v)) - 2}{\left(\frac{1}{D}(n-1)\right)^2 + \left(1 - \frac{1}{D}\right)d_G(u)d_G(v)} + \frac{(n-1)}{D}\left(1 - \frac{1}{D}\right)(d_G(u) + d_G(v))} \quad (4.14)$$

$$(ii) \sum_{uv \in E(G)} \left[\frac{(n-1)^2 + (n-1)(d_G(u) + d_G(v)) + d_G(u)d_G(v)}{2\left(2(n-1) + d_G(u) + d_G(v) - 4\right)}\right]^3$$

$$\leq RT_{AZ}(G) \leq$$

$$\sum_{uv \in E(G)} \left[\frac{\left(\frac{1}{D}(n-1)\right)^2 + \left(1 - \frac{1}{D}\right)d_G(u)d_G(v)}{+\frac{n-1}{D}\left(1 - \frac{1}{D}\right)(d_G(u) + d_G(v))} \middle/ \frac{2(n-1)}{D} + \left(1 - \frac{1}{D}\right)(d_G(u) + d_G(v)) - 2\right]^3, \quad (4.15)$$

$$(iii) \quad \sum_{uv \in E(G)} \frac{2(n-1) + (d_G(u) + d_G(v))}{2\sqrt{(n-1+d_G(u))(n-1+d_G(v))}} \quad \leq \quad RT_{AG}(G)$$

$$\leq \quad \sum_{uv \in E(G)} \frac{\frac{2(n-1)}{D} + \left(1 - \frac{1}{D}\right)(d_G(u) + d_G(v))}{2\sqrt{\begin{array}{l} \left(\frac{1}{D}(n-1)\right)^2 \\[4pt] + \frac{1}{D}(n-1)\left(1 - \frac{1}{D}\right)(d_G(u) + d_G(v)) \\[4pt] + \left(1 - \frac{1}{D}\right)^2 d_G(u)d_G(v) \end{array}}} \qquad (4.16)$$

and

$$(iv) \quad \sum_{uv \in E(G)} \frac{2\sqrt{(n-1+d_G(u))(n-1+d_G(v))}}{2(n-1) + (d_G(u) + d_G(v))} \quad \leq \quad RT_{GA}(G)$$

$$\leq \quad \sum_{uv \in E(G)} \frac{2\sqrt{\begin{array}{l} \left(\frac{1}{D}(n-1)\right)^2 \\[4pt] + \frac{1}{D}(n-1)\left(1 - \frac{1}{D}\right)(d_G(u) + d_G(v)) \\[4pt] + \left(1 - \frac{1}{D}\right)^2 d_G(u)d_G(v) \end{array}}}{\frac{2(n-1)}{D} + \left(1 - \frac{1}{D}\right)(d_G(u) + d_G(v))}. \qquad (4.17)$$

4.3 Status and Reciprocal Status Based Topological Indices of Some Graphs

For any vertex u of a complete graph K_n, we have $\sigma_G(u) = n - 1$. Hence we get the Proposition 4.3.1.

Proposition 4.3.1. *For a complete graph K_n on n vertices,*

$$T_{SC}(K_n) = \frac{n\sqrt{n-1}}{2\sqrt{2}},$$

$$T_{ABC}(K_n) = \frac{n\sqrt{n-2}}{\sqrt{2}},$$

$$T_{AG}(K_n) = \frac{n(n-1)}{2}$$

and

$$T_{AZ}(K_n) = \frac{n(n-1)^7}{16(n-2)^3}.$$

The complete bipartite graph $K_{p,q}$ has $n = p + q$ vertices and $m = pq$ edges. Also $diam(K_{p,q}) \leq 2$. Therefore by the equality part of Theorems 4.2.1 to 4.2.4 we get the Proposition 4.3.2.

Proposition 4.3.2. *For a complete bipartite graph $K_{p,q}$,*

$$T_{SC}(K_{p,q}) = \frac{pq}{\sqrt{3(p+q)-4}},$$

$$T_{ABC}(K_{p,q}) = pq\sqrt{\frac{3(p+q)-6}{4(p+q)^2 - (2p+2q+6)(p+q) - 4 + pq}},$$

$$T_{AG}(K_{p,q}) = pq\left(\frac{3(p+q)-4}{2\sqrt{(2p+q-2)(2q+p-2)}}\right)$$

and

$$T_{AZ}(K_{p,q}) = pq\left[\frac{4(p+q)^2 - (2p+2q+6)(p+q) - 4 + pq}{3(p+q)-6}\right]^3.$$

For any vertex u of a cycle C_n on $n \geq 3$ vertices, we have [51]

$$\sigma_G(u) = \begin{cases} 2\left[1 + 2 + \cdots + \frac{n-1}{2}\right] + \frac{n}{2} = \frac{n^2}{4}, & \text{if } n \text{ is even} \\ 2\left[1 + 2 + \cdots + \frac{n-1}{2}\right] = \frac{n^2-1}{4}, & \text{if } n \text{ is odd}. \end{cases}$$

Using this and by the definitions of status based topological indices, we have Proposition 4.3.3

Proposition 4.3.3. *For a cycle C_n on $n \geq 3$ vertices,*

$$T_{SC}(C_n) = \begin{cases} \sqrt{2}, & \text{if } n \text{ is even} \\ n\sqrt{\frac{2}{n^2-1}}, & \text{if } n \text{ is odd,} \end{cases}$$

$$T_{ABC}(C_n) = \begin{cases} \frac{\sqrt{8(n^2-4)}}{n}, & \text{if } n \text{ is even} \\ n\sqrt{\frac{8(n^2-5)}{(n^2-1)^2}}, & \text{if } n \text{ is odd,} \end{cases}$$

$$T_{AG}(C_n) = n, \ \forall \ n$$

and

$$T_{AZ}(C_n) = \begin{cases} n\left(\frac{n^4}{8(n^2-4)}\right)^3, & \text{if } n \text{ is even} \\ n\left(\frac{(n^2-1)^2}{8(n^2-5)}\right)^3, & \text{if } n \text{ is odd.} \end{cases}$$

Proposition 4.3.4. *For a wheel graph W_{n+1}, $n \geq 3$*

$$T_{SC}(W_{n+1}) = n\left(\frac{1}{\sqrt{3n-3}} + \frac{1}{\sqrt{4n-6}}\right),$$

$$T_{ABC}(W_{n+1}) = n\left[\sqrt{\frac{3n-5}{2n^2-3n-8}} + \sqrt{\frac{4n-8}{4n^2-12n+1}}\right],$$

$$T_{AG}(W_{n+1}) = n\left[\frac{3n-3}{2\sqrt{n(2n-3)}} + 1\right]$$

and

$$T_{AZ}(W_{n+1}) = n\left[\sqrt{\frac{2n^2-3n-8}{3n-5}} + \sqrt{\frac{4n^2-12n+1}{4n-8}}\right].$$

Proof. Partition the edge set $E(W_{n+1})$ into two sets E_1 and E_2, where

$$E_1 = \{uv \mid d_{W_{n+1}}(u) = n \text{ and } d_{W_{n+1}}(v) = 3\} \text{ and }$$

$E_2 = \{uv \mid d_{W_{n+1}}(u) = 3 \text{ and } d_{W_{n+1}}(v) = 3\}$. Also $|E_1| = n$ and $|E_2| = n$ and $diam(W_{n+1}) = 2$. Therefore by the equality part of Theorem 4.2.1, we get,

$$
\begin{aligned}
T_{SC}(W_{n+1}) &= \sum_{uv \in E(W_{n+1})} \frac{1}{\sqrt{4n - 4 - (d_{W_{n+1}}(u) + d_{W_{n+1}}(v))}} \\
&= \sum_{uv \in E_1} \frac{1}{\sqrt{4(n+1) - 4 - (n+3)}} \\
&\quad + \sum_{uv \in E_2} \frac{1}{\sqrt{4(n+1) - 4 - 6}} \\
&= \frac{n}{\sqrt{4n - n - 3}} + \frac{n}{\sqrt{4n - 6}}.
\end{aligned}
$$

Other results can be proved analogously. $\square$

Proposition 4.3.5. *For a friendship graph F_n, $n \geq 2$,*

$$T_{SC}(F_n) = \frac{2n}{\sqrt{6n - 2}} + \frac{n}{\sqrt{8n - 4}},$$

$$T_{ABC}(F_n) = 2n\sqrt{\frac{6n - 4}{24n^2 + 12n - 4}} + n\sqrt{\frac{8n - 6}{32n^2}},$$

$$T_{AG}(F_n) = 2n\left(\frac{6n - 2}{2\sqrt{2n(4n - 2)}}\right) + n$$

and

$$T_{AZ}(F_n) = 2n\left[\frac{8n^2 - 16n - 8}{6n - 4}\right]^3 + n\left[\frac{16n^2 + 8n + 4}{8n - 6}\right]^3.$$

Proof. The edge set of a graph F_n can be partitioned into two sets E_1 and E_2 such that, $E_1 = \{uv \mid d_{F_n}(u) = 2n \text{ and } d_{F_n}(v) = 2\}$ and $E_2 = \{uv \mid d_{F_n}(u) = 2 \text{ and } d_{F_n}(v) = 2\}$. Also $|E_1| = 2n$ and $|E_2| = n$ and $diam(F_n) = 2$. Therefore by the equality part of Theorem 4.2.1 we have,

$$
\begin{aligned}
T_{SC}(F_n) &= \sum_{uv \in E(G)} \frac{1}{\sqrt{4n - 4 - (d_{F_n}(u) + d_{F_n}(v))}} \\
&= \sum_{uv \in E_1} \frac{1}{\sqrt{4(2n+1) - 4 - (2n+2)}} \\
&\quad + \sum_{uv \in E_2} \frac{1}{\sqrt{4(2n+1) - 4 - 4}} \\
&= \frac{2n}{\sqrt{6n - 2}} + \frac{n}{8n - 4}.
\end{aligned}
$$

Other results can be proved analogously. $\qquad\square$

For any vertex u in K_n, we have $rs_{K_n}(u) = n - 1$.

Proposition 4.3.6. *For a complete graph K_n on n-vertices,*

$$
RT_{SC}(K_n) = \frac{n\sqrt{n-1}}{2\sqrt{2}},
$$

$$
RT_{ABC}(K_n) = \frac{n\sqrt{n-2}}{\sqrt{2}},
$$

$$
RT_{AZ}(K_n) = \frac{n(n-1)^7}{16(n-2)^3},
$$

$$
RT_{AG}(K_n) = \frac{n(n-1)}{2}
$$

and

$$RT_{GA}(K_n) = \frac{n(n-1)}{2}.$$

Proposition 4.3.7. *For a complete bipartite graph $K_{p,q}$,*

$$RT_{SC}(K_{p,q}) = \frac{pq}{\sqrt{\frac{1}{2}(p+q) - 1}}.$$

Proof. The vertex set $V(K_{p,q})$ can be partitioned into two independent sets V_1 and V_2 such that the vertex $u \in V_1$ and $v \in V_2$ for every $uv \in E(K_{p,q})$. Therefore $d_{K_{p,q}}(u) = q$ and $d_{K_{p,q}}(v) = p$. And the graph $K_{p,q}$ has $n = p + q$ vertices and $m = pq$ edges. Also $diam(G) \leq 2$. Therefore by the equality part of Theorem 4.2.5 we get,

$$RT_{SC}(K_{p,q}) = \sum_{uv \in E(G)} \frac{1}{\sqrt{(n-1) + \frac{1}{2}(d_{K_{p,q}}(u) + d_{K_{p,q}}(v))}}$$

$$RT_{SC}(K_{p,q}) = \sum_{uv \in E(G)} \sqrt{\frac{1}{p+q-1+\frac{1}{2}(p+q)}}$$

$$= \sum_{uv \in E(G)} \frac{1}{\sqrt{\frac{1}{2}(p+q) - 1}}$$

$$= \frac{pq}{\sqrt{\frac{1}{2}(p+q) - 1}}.$$

$\square$

By the equality part of Eqs. (4.14) to (4.17) we have following.

Proposition 4.3.8. *For a complete bipartite graph $K_{p,q}$,*

$$RT_{ABC}(K_{p,q}) = pq\sqrt{\frac{6(p+q) - 12}{(p+q-1)^2 + (p+q)(p+q-1) + pq}},$$

$$RT_{AZ}(K_{p,q}) = pq \left[\frac{(p+q-1)^2 + (p+q)(p+q-1) + pq}{6(p+q) - 12} \right]^3,$$

$$RT_{AG}(K_{p,q}) = pq \left(\frac{p+q-2}{2\sqrt{(p+q-1)^2 + (p+q)(p+q-1) + pq}} \right)$$

and

$$RT_{GA}(K_{p,q}) = \left(\frac{2\sqrt{(p+q-1)^2 + (p+q)(p+q-1) + pq}}{p+q-2} \right).$$

Proposition 4.3.9. *For a cycle C_n on $n \geq 3$ vertices,*

$$RT_{SC}(C_n) = \begin{cases} \dfrac{n}{2\sqrt{\sum_{i=1}^{\frac{n-2}{2}} \frac{1}{i} + \frac{1}{n}}}, & \text{if} \quad n \text{ is even} \\[3em] \dfrac{n}{2\sqrt{\sum_{i=1}^{\frac{n-1}{2}} \frac{1}{i}}}, & \text{if} \quad n \text{ is odd.} \end{cases}$$

Proof. If n is even number then for any vertex $u \in V(C_n)$,

$$rs_{C_n}(u) = 2 \sum_{i=1}^{\frac{n-2}{2}} \frac{1}{i} + \frac{2}{n}.$$

Therefore

$$\begin{aligned} RT_{SC}(C_n) &= \sum_{uv \in E(G)} \frac{1}{\sqrt{rs_{C_n}(u) + rs_{C_n}(v)}} \\ &= \sum_{uv \in E(G)} \frac{n}{2\sqrt{\sum_{i=1}^{\frac{n-2}{2}} \frac{1}{i} + \frac{1}{n}}}. \end{aligned}$$

If n is odd number then for any vertex $u \in V(C_n)$,

$$rs_{C_n}(u) = 2 \sum_{i=1}^{\frac{n-1}{2}} \frac{1}{i}.$$

Therefore

$$RT_{SC}(C_n) = \sum_{uv \in E(G)} \frac{1}{\sqrt{4\sum_{i=1}^{\frac{n-1}{2}} \frac{1}{i}}}$$

$$= \frac{n}{2\sqrt{\sum_{i=1}^{\frac{n-1}{2}} \frac{1}{i}}}.$$

$\square$

Proposition 4.3.10. *For a cycle C_n on $n \geq 3$ vertices,*

$$RT_{ABC}(C_n) = \begin{cases} n\dfrac{\sqrt{4\left(\sum_{i=1}^{\frac{n-2}{2}} \frac{1}{i}+\frac{1}{n}\right)-2}}{2\left(\sum_{i=1}^{\frac{n-2}{2}} \frac{1}{i}+\frac{1}{n}\right)}, & \text{if } n \text{ is even,} \\[4ex] n\dfrac{\sqrt{4\sum_{i=1}^{\frac{n-1}{2}} \frac{1}{i}-2}}{2\sum_{i=1}^{\frac{n-1}{2}} \frac{1}{i}} & \text{if } n \text{ is odd,} \end{cases}$$

$$RT_{AZ}(C_n) = \begin{cases} n\left(\dfrac{2\left(\sum_{i=1}^{\frac{n-2}{2}} \frac{1}{i}+\frac{1}{n}\right)}{\sqrt{4\left(\sum_{i=1}^{\frac{n-2}{2}} \frac{1}{i}+\frac{1}{n}\right)-2}}\right)^3, & \text{if } n \text{ is even,} \\[4ex] n\left(\dfrac{2\sum_{i=1}^{\frac{n-1}{2}} \frac{1}{i}}{\sqrt{4\sum_{i=1}^{\frac{n-1}{2}} \frac{1}{i}-2}}\right)^3 & \text{if } n \text{ is odd,} \end{cases}$$

$$RT_{AG}(C_n) = n, \quad for \ all \ n$$

and

$$RT_{GA}(C_n) = n, \quad for \ all \ n.$$

Proposition 4.3.11. *For a wheel W_{n+1}, $n \geq 3$,*

$$RT_{SC}(W_{n+1}) = \frac{n}{\sqrt{n+\frac{1}{2}(n+3)}} + \frac{n}{\sqrt{n+3}}.$$

Proof. Here we partition the edge set $E(W_{n+1})$ into two sets E_1 and E_2, where $E_1 = \{uv | d_{W_{n+1}}(u) = n \text{ and } d_{W_{n+1}}(v) = 3\}$ and $E_2 = \{uv | d_{W_{n+1}}(u) = 3 \text{ and } d_{W_{n+1}}(v) = 3.\}$ Also $|E_1| = n$ and $|E_2| = n$ and $diam(W_{n+1}) = 2$. Therefore by the equality part of Theorem 4.2.5 , we get,

$$
\begin{aligned}
RT_{SC}(W_{n+1}) &= \sum_{uv \in E(G)} \frac{1}{\sqrt{(n-1) + \frac{1}{2}(d_{W_{n+1}}(u) + d_{W_{n+1}}(v))}} \\
&= \sum_{uv \in E_1} \frac{1}{\sqrt{(n-1) + \frac{1}{2}(d_{W_{n+1}}(u) + d_{W_{n+1}}(v))}} \\
&\quad + \sum_{uv \in E_2} \frac{1}{\sqrt{(n-1) + \frac{1}{2}(d_{W_{n+1}}(u) + d_{W_{n+1}}(v))}} \\
&= \sum_{uv \in E_1} \frac{1}{\sqrt{n + \frac{1}{2}(n+3)}} + \sum_{uv \in E_2} \frac{1}{\sqrt{n+3}} \\
&= \frac{1}{\sqrt{(n-1) + \frac{1}{2}(d_{W_{n+1}}(u) + d_{W_{n+1}}(v))}} \\
&= \frac{n}{\sqrt{n + \frac{1}{2}(n+3)}} + \frac{n}{\sqrt{n+3}}.
\end{aligned}
$$

$\square$

Proposition 4.3.12. *For a wheel W_{n+1}, $n \geq 3$,*

$$
RT_{ABC}(W_{n+1}) = n\sqrt{\frac{2(3n-1)}{2n^2 + 6n}} + n\sqrt{\frac{2(2n+2)}{n^2 + 6n + 9}},
$$

$$
RT_{AZ}(W_{n+1}) = n\left[\frac{2n^2 + 6n}{2(3n-1)}\right]^3 + n\left[\frac{n^2 + 6n + 9}{2(2n+2)}\right]^3,
$$

$$
RT_{AG}(W_{n+1}) = \frac{n(3n+3)}{2\sqrt{2n^2 + 6n}} + \frac{n(2n+6)}{2\sqrt{n^2 + 6n + 9}}
$$

and

$$RT_{GA}(W_{n+1}) = \frac{2n\sqrt{2n^2 + 6n}}{3n + 3} + \frac{2n\sqrt{n^2 + 6n + 9}}{2n + 6}.$$

Proposition 4.3.13. *For a friendship graph F_n, $n \geq 2$,*

$$RT_{SC}(F_n) = \frac{2n}{\sqrt{3n + 1}} + \frac{n}{\sqrt{2n + 2}}.$$

Proof. The edge set of a graph F_n can be partitioned into two sets E_1 and E_2 such that, $E_1 = \{uv | d_{F_n}(u) = 2n \text{ and } d_{F_n}(v) = 2\}$ and $E_2 = \{uv | d_{F_n}(u) = 2 \text{ and } d_{F_n}(v) = 2.\}$ Also $|E_1| = 2n$ and $|E_2| = n$ and $diam(F_n) = 2$. Therefore by the equality part of Theorem 4.2.5 we have,

$$
\begin{aligned}
RT_{SC}(F_n) &= \sum_{uv \in E(F_n)} \frac{1}{\sqrt{n - 1 + \frac{1}{2}(d_{F_n}(u) + d_{F_n}(v))}} \\
&= \sum_{uv \in E_1} \frac{1}{\sqrt{n - 1 + \frac{1}{2}(d_{F_n}(u) + d_{F_n}(v))}} \\
&\quad + \sum_{uv \in E_2} \frac{1}{\sqrt{n - 1 + \frac{1}{2}(d_{F_n}(u) + d_{F_n}(v))}} \\
&= \sum_{uv \in E_1} \frac{1}{\sqrt{2n + n + 1}} + \sum_{uv \in E_2} \frac{1}{\sqrt{2n + 2}} \\
&= \frac{2n}{\sqrt{3n + 1}} + \frac{n}{\sqrt{2n + 2}}.
\end{aligned}
$$

$\square$

Proposition 4.3.14. *For a friendship graph F_n, $n \geq 2$,*

$$RT_{ABC}(F_n) = 2n \left(\sqrt{\frac{3n - 1}{2n^2 + 2n}} \right) + n \left(\sqrt{\frac{4n}{n^2 + 2n + 1}} \right),$$

$$RT_{AZ}(F_n) = 2n \left[\frac{2n^2 + 2n}{3n - 1}\right]^3 + n \left[\frac{n^2 + 2n + 1}{4n}\right]^3,$$

$$RT_{AG}(F_n) = \frac{n(6n + 2)}{\sqrt{8n^2 + 8n}} + \frac{n(2n + 2)}{\sqrt{4n^2 + 8n + 4}}$$

and

$$RT_{GA}(F_n) = \frac{2n\sqrt{8n^2 + 8n}}{3n + 1} + \frac{n\sqrt{n^2 + 2n + 1}}{n + 1}.$$

4.4 Regression Analysis Using Status and Reciprocal Status Based Topological Indices of Graphs

In this section we discuss the correlation between status based topological indices and chemical properties such as boiling point (BP), density (DEN), molecular mass (MM) and melting point (MP) of paraffins. The correlation between the boiling point and status geometric-arithmetic index of paraffin is reported in [42].

In Table 4.1 the chemical properties of paraffins under consideration are given. In Table 4.2 the status based topological indices of molecular graph of paraffins are listed.

The scatter plot between the status based topological indices and chemical properties of paraffins are shown in Figs. 4.2 to 4.5.

Using the data of Tables 4.1 and 4.2, the linear regression models for the boiling point (BP), molecular mass (MM), density (DEN)

Status Like Topological Indices of Graphs and its Regression Analysis with Some Molecular Properties

Table 4.1: Paraffins and its chemical properties

Sl No.	Paraffins	Boiling point in deg cel	Density in mg/mL	Molecular mass in g/mol	Melting point in deg cel
1	2-methylpentane	62.9	653	86.18	-160
2	2,2-dimethylbutane	50	649	86.17	-102
3	2,3-dimethylbutane	57.9	662	86.18	-136
4	2,2-dimethylpentane	79	674	100.2	-102
5	3,3-dimethylpentane	86	693	100.2	-135
6	n-octane	125	703	114.23	-57
7	3-methylpentane	118	664	86.2	-162.8
8	3-ethylhexane	118	718	114.23	–
9	2,2-dimethylhexane	107	693	114.23	-121.1
10	2,4-dimethylhexane	108	701	114.23	-91.46
11	2-methyl, 3-ethylpentane	116	718	114.23	–
12	2,2,4-trimethylpentane	99	692	114.22	-107.38
13	n-dodecane	216.2	0.7495	170.34	-10

Table 4.2: Status based topological indices of paraffins

Sl No.	Paraffins	T_{GA}	T_{AG}	T_{SC}	T_{ABC}	T_{AZ}
1	2-methylpentane	4.924	5.07965	1.1621	1.7975	694.1082
2	2,2-dimethylbutane	4.872	5.1317	1.2408	1.3908	450.627
3	2,3-dimethylbutane	4.872	8.1048	1.21015	2.2819	513.9
4	2,2-dimethylpentane	5.903	6.0962	1.2354	2.40839	1558.7
5	3,3-dimethylpentane	5.880	6.1199	1.3103	2.4799	1302.501
6	n-octane	6.968	7.0381	1.1188	1.7539	6456.095
7	3-methylpentane	6.946	6.05026	1.1758	2.314	5941.7
8	3-ethylhexane	6.930	7.067	1.2283	2.8383	4864.35
9	2,2-dimethylhexane	6.926	7.0766	1.2488	2.4381	4337.283
10	2,4-dimethylhexane	6.930	7.1491	1.2706	2.4993	3671.56
11	2-methyl, 3-ethylpentane	6.908	8.1077	1.2819	2.0794	3626.635
12	2,2,4-trimethylpentane	6.902	7.0965	1.2903	2.52	3338.213
13	n-dodecane	10.974	15.73	1.16175	2.97226	97907.83

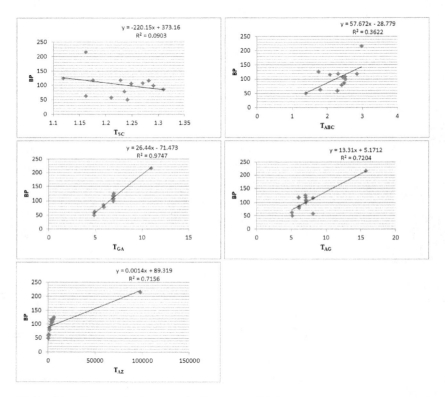

Figure 4.2 : Scatter plot between boiling point (BP) and status based topological indices.

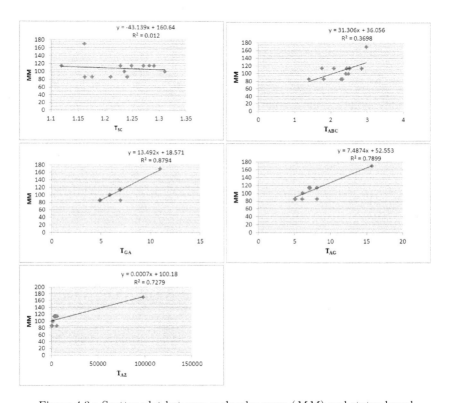

Figure 4.3 : Scatter plot between molecular mass (MM) and status based topological indices.

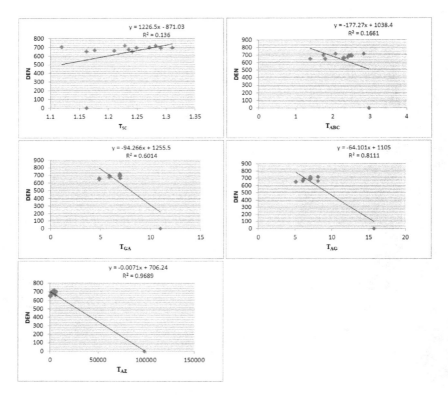

Figure 4.4 :Scatter plot between density (DEN) and status based topological indices.

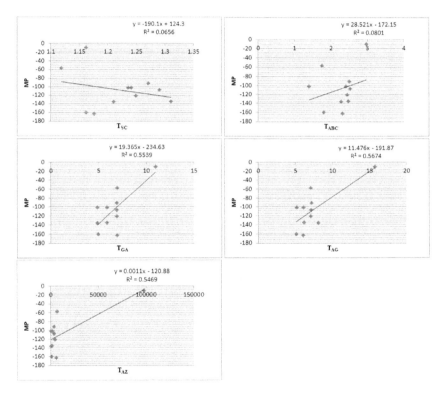

Figure 4.5 :Scatter plot between melting point (MP) and status based topological indices.

Table 4.3: Correlation coefficient between status based topological indices and chemical properties of paraffins

Index	BP	MM	DEN	MP
T_{SC}	0.3	0.11	0.369	0.256
T_{ABC}	0.602	0.608	0.408	0.283
T_{GA}	0.986	0.937	0.775	0.743
T_{AG}	0.84	0.889	0.901	0.753
T_{AZ}	0.846	0.853	0.984	0.74

and melting point (MP) of paraffins in terms of the status based topological indices are obtained using least square fitting and they are represented by the Eqs. (4.18) to (4.37).

In Table 4.3 the correlation coefficients between status based topological indices and chemical properties of paraffins are listed, where as in the Table 4.4 the standard error of estimation between status based topological indices and chemical properties of paraffins are listed.

$$BP = 373.156(\pm 258.523) - 220.146(\pm 210.693)T_{SC} \quad (4.18)$$

$$BP = -28.779(\pm 53.738) + 57.672(\pm 23.074)T_{ABC} \quad (4.19)$$

$$BP = -71.47(\pm 8.715) + 26.44(\pm 1.285)T_{GA} \quad (4.20)$$

$$BP = 5.171(\pm 19.529) + 13.310(\pm 2.5)T_{AG} \quad (4.21)$$

$$BP = 89.319(\pm 7.0291) + 0.001(\pm 0.000)T_{AZ} \quad (4.22)$$

Status Like Topological Indices of Graphs and its Regression Analysis with Some Molecular Properties

Table 4.4: Standard error of estimation between status based topological indices and chemical properties of paraffins

Index	BP	MM	DEN	MP
T_{SC}	41.9598	23.491	185.587	45.50181
T_{ABC}	35.133	18.7611	182.3276	45.14748
T_{GA}	7.002	8.2086	126.052	31.4388
T_{AG}	23.2612	10.83293	86.7742	30.9845
T_{AZ}	23.4608	12.32818	35.2393	31.68549

$$MM = 160.635(\pm 144.734) - 43.139(\pm 117.956)T_{SC} \quad (4.23)$$

$$MM = 36.056(\pm 28.696) + 31.306(\pm 12.322)T_{ABC} \quad (4.24)$$

$$MM = 18.57(\pm 10.217) + 13.49(\pm 1.507)T_{GA} \quad (4.25)$$

$$MM = 52.553(\pm 9.095) + 7.487(\pm 1.164)T_{AG} \quad (4.26)$$

$$MM = 100.178(3.694) + 0.001(\pm 0.000)T_{AZ} \quad (4.27)$$

$$DEN = -871.029(\pm 1143.442) + 1226.491(\pm 931.891)T_{SC} \quad (4.28)$$

$$DEN = 1038.363(\pm 278.878) - 177.269(\pm 119.747)T_{ABC} \quad (4.29)$$

$$DEN = 1255(\pm 156.892) - 94.26(\pm 23.137)T_{GA} \quad (4.30)$$

$$DEN = 1104(\pm 72.852) - 64.101(\pm 9.326)T_{AG} \quad (4.31)$$

$$DEN = 706.242(\pm 10.558) - 0.007(\pm 0.000)T_{AZ} \quad (4.32)$$

$$MP = 124.304(\pm292.227) - 190.102(\pm239.181)T_{SC} \quad (4.33)$$

$$MP = -172.151(\pm74.071) + 28.521(\pm32.222)T_{ABC} \quad (4.34)$$

$$MP = -234.6(\pm39.133) + 19.36(\pm5.793)T_{GA} \quad (4.35)$$

$$MP = -191.869(\pm26.213) + 11.476(\pm3.340)T_{AG} \quad (4.36)$$

$$MP = -120.880(\pm10.356) + 0.001(\pm0.00)T_{AZ} \quad (4.37)$$

Observing the values of Table 4.3 we see that the regression model given in Eq. (4.20) gives the good correlation ($R = 0.986$) of status geometric-arithmetic index with boiling point of paraffins compared to other status based topological indices considered in this chapter. It was reported in [42].

The regression model given in Eq. (4.25) gives the good correlation ($R = 0.937$) of status geometric-arithmetic index with molecular mass of paraffins compared to other status based topological indices.

The regression model given in Eq. (4.32) gives the good correlation ($R = 0.984$) of status augmented Zagreb index with density of paraffins compared to other status based topological indices.

The regression model given in Eq. (4.36) gives good correlation ($R = 0.753$) of status arithmetic-geometric index with melting point of paraffins compared to other status based topological indices considered in this chapter.

Status sum-connectivity index and status atom-bond connectivity index has poor correlation with the chemical properties of paraffins.

Now we discuss the correlation between reciprocal status based topological indices and chemical properties such as boiling point (BP), density (DEN), molecular mass (MM) and melting point (MP) of paraffins.

In Table 4.1 the chemical properties of paraffins under consideration are given. In Table 4.5 the reciprocal status based topological indices of molecular graph of paraffins are listed.

The scatter plot between the reciprocal status based topological indices and chemical properties of paraffins are shown in Figs. 4.6 to 4.9.

Using the data of Table 4.1 and Table 4.5, the linear regression models for the boiling point (BP), molecular mass (MM), density (DEN) and melting point (MP) of paraffins in terms of the reciprocal status based topological indices are obtained using least square fitting and they are represented by the Eqs. (4.38) to (4.57).

In Table 4.6 the correlation coefficients between reciprocal status based topological indices and chemical properties of paraffins are listed, where as in the Table 4.7 the standard error of estimation between reciprocal status based topological indices and chemical properties of paraffins are listed.

Table 4.5: Reciprocal status based topological indices of paraffins

Sl No.	RT_{AG}	RT_{GA}	RT_{SC}	RT_{ABC}	RT_{AZ}
1	5.0555	4.9447	1.9609	3.2919	62.8403
2	2.3019	4.8961	1.8711	3.2338	69.475
3	5.0846	4.91819	1.905	3.2523	30.642
4	6.09462	5.9315	2.1726	3.70189	97.8068
5	6.0977	5.9033	2.1524	3.7602	102.295
6	7.0225	6.9761	2.6363	4.4678	105.3818
7	7.04472	6.9488	2.5466	4.3936	117.335
8	7.0516	6.9452	2.5285	4.3692	122.8227
9	7.0766	6.92612	2.4835	4.3023	134.492
10	7.06192	6.9363	2.5097	4.3577	123.6375
11	7.07975	6.9134	2.4678	4.3092	133.014
12	7.2225	6.7627	2.4518	4.3952	122.098
13	10.02417	10.9741	3.7552	6.5911	244.79

Table 4.6: Correlation coefficient between reciprocal status based topological indices and chemical properties of paraffins

Index	BP	MM	DEN	MP
RT_{SC}	0.994	0.927	0.778	0.751
RT_{ABC}	0.989	0.933	0.775	0.746
RT_{AZ}	0.944	0.923	0.731	0.702
RT_{AG}	0.901	0.820	0.533	0.522
RT_{GA}	0.989	0.937	0.781	0.747

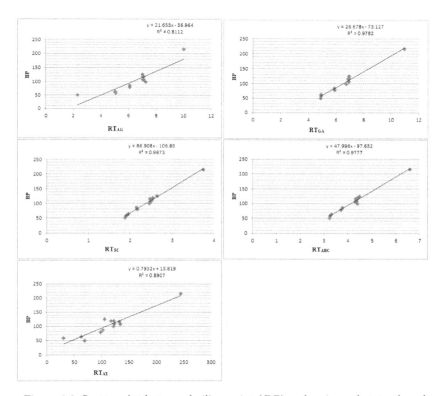

Figure 4.6 :Scatter plot between boiling point (BP) and reciprocal status based topological indices.

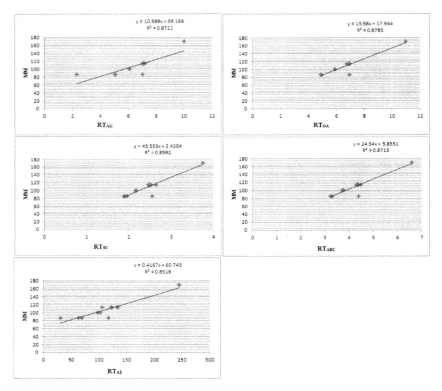

Figure 4.7 :Scatter plot between molecular mass (MM) and reciprocal status based topological indices.

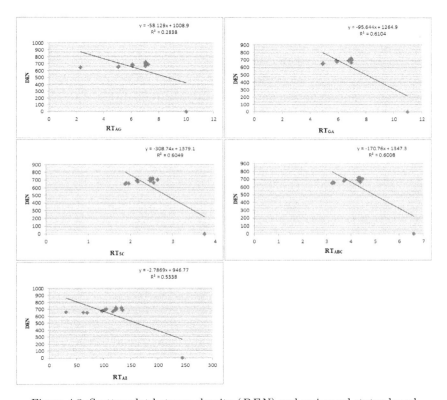

Figure 4.8 :Scatter plot between density (DEN) and reciprocal status based topological indices.

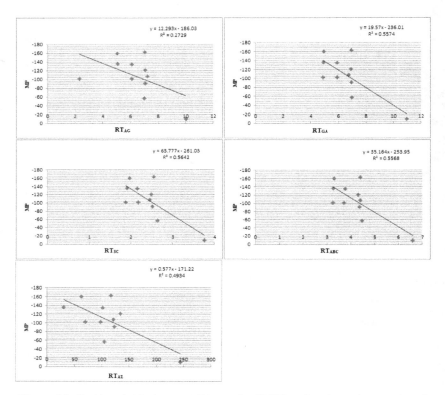

Figure 4.9 :Scatter plot between melting point (MP) and reciprocal status based topological indices.

Table 4.7: Standard error of estimation between reciprocal status based topological indices and chemical properties of paraffins

Index	BP	MM	DEN	MP
RT_{SC}	4.9599	8.8693	125.5084	31.07494
RT_{ABC}	6.5625	8.47825	126.1470	31.3383
RT_{AZ}	14.5424	9.09560	136.3334	33.5049
RT_{AG}	19.1169	13.5338	168.9718	40.13722
RT_{GA}	6.4964	8.24441	124.6304	31.3154

Using the data of Tables 4.1 and 4.5, the linear regression models for the boiling point (BP), molecular mass (MM), density (DEN) and melting point (MP) of paraffins interms of the reciprocal satus based topological indices are obtained using least square fitting and they are represented by the Eqs. (4.38) to (4.57).

$$BP = -106.8859(\pm 7.321) + 86.908(\pm 2.973)RT_{SC} \quad (4.38)$$

$$BP = -97.632(\pm 9.319) + 47.996(\pm 2.183)RT_{ABC} \quad (4.39)$$

$$BP = 13.819(\pm 10.275) + 0.793(\pm 0.084)RT_{AZ} \quad (4.40)$$

$$BP = -36.964(\pm 21.083) + 21.653(\pm 3.150)RT_{AG} \quad (4.41)$$

$$BP = -73.127(\pm 8.145) + 26.678(\pm 1.201)RT_{GA} \quad (4.42)$$

$$DEN = 1379.069(\pm185.266) - 308.738(\pm75.237)RT_{SC} \quad (4.43)$$

$$DEN = 1347.2789(\pm179.142) - 170.761(\pm41.965)RT_{ABC} \quad (4.44)$$

$$DEN = 946.773(\pm96.328) - 2.787(\pm0.785)RT_{AZ} \quad (4.45)$$

$$DEN = 1008.945(\pm186.354) - 58.129(\pm27.841)RT_{AG} \quad (4.46)$$

$$DEN = 1264.912(\pm156.248) - 95.644(\pm23.040)RT_{GA} \quad (4.47)$$

$$MM = 2.420(\pm13.092) + 43.553(\pm5.317)RT_{SC} \quad (4.48)$$

$$MM = 5.855(\pm12.040) + 24.340(\pm2.821)RT_{ABC} \quad (4.49)$$

$$MM = 60.743(\pm6.427) + 0.417(\pm0.052)RT_{AZ} \quad (4.50)$$

$$MM = 39.166(\pm14.926) + 10.588(\pm2.230)RT_{AG} \quad (4.51)$$

$$MM = 17.944(10.336) + 13.580(\pm1.524)RT_{GA} \quad (4.52)$$

$$MP = -261.031(\pm45.887) + 63.777(\pm18.685)RT_{SC} \quad (4.53)$$

$$MP = -253.946(\pm44.509) + 35.164(\pm10.458)RT_{ABC} \quad (4.54)$$

$$MP = -171.219(\pm230.714) + 0.577(\pm0.195)RT_{AZ} \quad (4.55)$$

$$MP = -186.026(\pm44.297) + 12.293(\pm6.688)RT_{AG} \quad (4.56)$$

$$MP = -236.0069)(\pm39.261) + 19.570(\pm5.813)RT_{GA} \quad (4.57)$$

Observing the values of Table 4.6 we see that the regression model given in Eq. (4.38) gives the good correlation ($R = 0.994$) of reciprocal status sum-connectivity index with boiling point of paraffins compared to other reciprocal status based topological indices considered in this chapter.

The regression model given in Eq. (4.47) gives the good correlation ($R = 0.937$) of reciproal status geometric-arithmetic index with molecular mass of paraffins compared to other reciprocal status based topological indices.

The regression model given in Eq. (4.52) gives the good correlation ($R = 0.781$) of reciproal status geometric-arithmetic index with density of paraffins compared to other reciprocal status based topological indices.

The regression model given in Eq. (4.53) gives the good correlation ($R = 0.751$) of reciprocal status sum-connectivity index with molecular mass of paraffins compared to other reciprocal status based topological indices considered in this chapter.

From the observation it is clear that the reciprocal status based topological indices has good correlation with the chemical properties of paraffins compared to status based topological indices.

Chapter 5

Zagreb Indices and Co-indices of Total graph, Semi-total Point Graph and Semi-total Line Graph of Subdivision Graphs

---

Results of this chapter are published in:

H. S. Ramane, S. Y. Talwar, I. Gutman "Zagreb indices and co-indices of total graph, semi-total point graph and semi-total line graph of subdivision graphs", Mathematics Interdisciplinary Research, 5 (1) (2020) 1-12 [ISSN: 2538 3639].

5.1 Introduction

The Zagreb indices belongs among the oldest molecular structure descriptors based on degrees of vertices and these are extensively studied by many researchers [43].

Ranjini et al. [54] calculated the Zagreb indices and co-indices of the line graph of the subdivision graph of tadpole graph, wheel and ladder. Later Ramane et al. [48] generalized these results by finding the Zagreb indices of the line graph of subdivision graph of any graph.

In [40], Mohanappriya and Vijayalakshmi obtained the Zagreb indices of the total graph of subdivision graph of tadpole graph, wheel and ladder.

In this chapter we obtain the Zagreb indices and co-indices of the total graph of the subdivision graph of any graph. Thus generalizes the results of Mohanappriya and Vijayalakshmi [40]. Further we compute the Zagreb indices of semi-total point graph and semi-total line graph of subdivision graph of any graph in terms of the parameters of underline graph.

The subdivision graph $S(G)$ is the graph obtained from G by inserting a new vertex into each edge of G. The tadpole $T_{n,k}$ is the graph obtained by joining one vertex of cycle C_n to the one end point of path P_k. The wheel W_{n+1} is the graph obtained by joining all vertices of C_n to the new vertex.

The Cartesian product $G_1 \times G_2$ of G_1 and G_2 is a graph with vertex set $V(G_1) \times V(G_2)$ and two vertices (u_1, v_1) and (u_2, v_2) are adjacent in $G_1 \times G_2$ if and only if either $u_1 = u_2$ and v_1 is adjacent to v_2 in G_2 or $v_1 = v_2$ and u_1 is adjacent to u_2 in G_1. The Ladder L_n is given by $L_n = K_2 \times P_n$, where P_n is a path on n vertices and K_2 is the complete graph on 2 vertices.

5.2 Total Graph, Semi-total Point Graph and Semi-total Line Graph

The vertices and edges of G are referred as their elements. The total graph of G, denoted by $T(G)$, is a graph with vertex set $V(T(G)) = V(G) \cup E(G)$ and two vertices in $T(G)$ are adjacent if and only if they are adjacent elements or they are incident elements in G [28].

The semi-total point graph of G, denoted by $T_1(G)$, is a graph with vertex set $V(T_1(G)) = V(G) \cup E(G)$ and two vertices in $T_1(G)$ are adjacent if they are adjacent vertices in G or one is vertex and other is an edge, incident to it [56].

The semi-total line graph of G, denoted by $T_2(G)$, is a graph with vertex set $V(T_2(G)) = V(G) \cup E(G)$ and two vertices in $T_2(G)$ are adjacent if they are adjacent edges in G or one is vertex and other is an edge, incident to it.

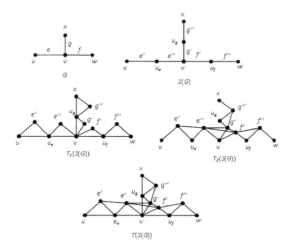

Figure 5.1 : Graph G, $S(G)$, $T_1(S(G))$, $T_2(S(G))$ and $T(S(G))$.

Observation 5.2.1. *If $u \in V(G)$, then $d_{S(G)}(u) = d_G(u)$ and if v is a subdivision vertex, then $d_{S(G)}(v) = 2$.*

Observation 5.2.2. *If $u \in V(G)$, then $d_{T(G)}(u) = 2d_G(u)$, $d_{T_1(G)}(u) = 2d_G(u)$ and $d_{T_2(G)}(u) = d_G(u)$.*

Observation 5.2.3. *If $e = uv \in E(G)$, then $d_{T(G)}(e) = d_G(u) + d_G(v)$, $d_{T_1(G)}(e) = 2$ and $d_{T_2(G)}(e) = d_G(u) + d_G(v)$.*

Without loss of generality, referring to the Fig. 5.1, let e and f be adjacent edges at v in G. Let e' and e'' be the subdivision edges of an edge e in $S(G)$ and f' and f'' be the subdivision edges of an edge f in $S(G)$. Let u_e and u_f be the subdivision vertices on edges e and f respectively in $S(G)$ (see Fig. 5.1).

Edge set $E(T(S(G)))$ of the total graph of the subdivision graph can be partitioned into sets E_1, E_2, E_3, E_4 and E_5, so that

$$E_1 = \{uu_e\}$$

where $u \in V(G)$ and u_e is the subdivision vertex in $S(G)$,

$$E_2 = \{ue'\}$$

where $u \in V(G)$ and e' is the subdivision edge in $S(G)$,

$$E_3 = \{u_e e'\}$$

where u_e is the subdivision vertex and e' is the subdivision edge in $S(G)$,

$$E_4 = \{e'e''\}$$

where e' and e'' are subdivision edges with common end vertex u_e in $S(G)$ and

$$E_5 = \{e''f'\}$$

where e'' and f' are subdivision edges with common end vertex v in $S(G)$ where $v \in V(G)$.

Easily we check that, $|E_1| = 2m$, $|E_2| = 2m$, $|E_3| = 2m$, $|E_4| = m$ and $|E_5| = \sum_{v \in V(G)} \frac{d_G(v)(d_G(v)-1)}{2} = -m + \frac{1}{2} \sum_{v \in V(G)} (d_G(v))^2$.

Observation 5.2.4. If $u \in V(G)$, then $d_{T(S(G))}(u) = 2d_G(u)$, $d_{T_1(S(G))}(u) = 2d_G(u)$ and $d_{T_2(S(G))}(u) = d_G(u)$.

Observation 5.2.5. If u_e is a subdivision vertex in $S(G)$, then $d_{T(S(G))}(u_e) = 4$, $d_{T_1(S(G))}(u_e) = 4$ and $d_{T_2(S(G))}(u_e) = 2$.

Observation 5.2.6. *If e' is a subdivision edge with one end vertex $u \in V(G)$, then $d_{T(S(G))}(e') = 2 + d_G(u)$, $d_{T_1(S(G))}(e') = 2$ and $d_{T_2(S(G))}(e') = 2 + d_G(u)$.*

5.3 Zagreb Indices

Theorem 5.3.1. *Let G be a graph with n vertices, m edges and vertex set $V(G)$. Then,*

$$(i) \ Z_1(T(S(G))) = 24m + 8Z_1(G) + \sum_{v \in V(G)} (d_G(v))^3,$$

$$(ii) \ Z_2(T(S(G))) = 16m + 18Z_1(G) + Z_2(G) + \frac{7}{2} \sum_{v \in V(G)} (d_G(v))^3$$

$$+ \frac{1}{2} \sum_{v \in V(G)} (d_G(v))^4.$$

Proof. By referring Fig. 5.1 and by Observations 5.2.4 to 5.2.6 we have

$$(i) \ Z_1(T(S(G))) = \sum_{uv \in E(T(S(G)))} \left[d_{T(S(G))}(u) + d_{T(S(G))}(v) \right]$$

$$= \sum_{uu_e \in E_1} \left[d_{T(S(G))}(u) + d_{T(S(G))}(u_e) \right]$$

$$+ \sum_{ue' \in E_2} \left[d_{T(S(G))}(u) + d_{T(S(G))}(e') \right]$$

$$+ \sum_{u-ee' \in E_3} \left[d_{T(S(G))}(u_e) + d_{T(S(G))}(e') \right]$$

$$+ \sum_{e'e'' \in E_4} \left[d_{T(S(G))}(e') + d_{T(S(G))}(e'') \right]$$

$$+ \sum_{e''f' \in E_5} \left[d_{T(S(G))}(e'') + d_{T(S(G))}(f') \right]$$

$$= \sum_{uv \in E(G)} \left[2d_G(u) + 4 + 2d_G(v) + 4 \right]$$

$$+ \sum_{uv \in E(G)} \left[2d_G(u) + 2 + d_G(u) + 2d_G(v) + 2 + d_G(v) \right]$$

$$+ \sum_{uv \in E(G)} \left[4 + 2 + d_G(u) + 4 + 2 + d_G(v) \right]$$

$$+ \sum_{uv \in E(G)} \left[2 + d_G(u) + 2 + d_G(v) \right]$$

$$+ \sum_{e''f' \in E_5} \left[2 + d_G(v) + 2 + d_G(v) \right]$$

$$= \left[8m + 2Z_1(G) \right] + \left[4m + 3Z_1(G) \right] + \left[12m + Z_1(G) \right]$$

$$+ \left[4m + Z_1(G) \right]$$

$$+ \sum_{v \in V(G)} \left[4 + 2d_G(v) \right] \left[\frac{d_G(v)(d_G(v) - 1)}{2} \right]$$

$$= 24m + 8Z_1(G) + \sum_{v \in V(G)} (d_G(v))^3.$$

$$(ii) \ \ Z_2(T(S(G))) = \sum_{uv \in E(T(S(G)))} \left(d_{T(S(G))}(u) \right) \left(d_{T(S(G))}(v) \right)$$

$$\begin{aligned}
&= \sum_{uu_e \in E_1} \left(d_{T(S(G))}(u)\right)\left(d_{T(S(G))}(u_e)\right) \\
&\quad + \sum_{ue' \in E_2} \left(d_{T(S(G))}(u)\right)\left(d_{T(S(G))}(e')\right) \\
&\quad + \sum_{u_e e' \in E_3} \left(d_{T(S(G))}(u_e)\right)\left(d_{T(S(G))}(e')\right) \\
&\quad + \sum_{e' e'' \in E_4} \left(d_{T(S(G))}(e')\right)\left(d_{T(S(G))}(e'')\right) \\
&\quad + \sum_{e'' f' \in E_5} \left(d_{T(S(G))}(e'')\right)\left(d_{T(S(G))}(f')\right) \\
&= \sum_{uv \in E(G)} [(2d_G(u))(4) + (2d_G(v))(4)] \\
&\quad + \sum_{uv \in E(G)} [(2d_G(u))(2 + d_G(u)) + (2d_G(v))(2 + d_G(v))] \\
&\quad + \sum_{uv \in E(G)} [(4)(2 + d_G(u)) + (4)(2 + d_G(v))] \\
&\quad + \sum_{uv \in E(G)} [(2 + d_G(u))(2 + d_G(v))] \\
&\quad + \sum_{e'' f' \in E_5} [(2 + d_G(v))(2 + d_G(v))] \\
&= [8Z_1(G)] + \left[4Z_1(G) + 2\sum_{uv \in E(G)} \left[(d_G(u))^2 + (d_G(v))^2\right]\right] \\
&\quad + [16m + 4Z_1(G)] \\
&\quad + [4m + 2Z_1(G) + Z_2(G)] \\
&\quad + \sum_{v \in V(G)} (2 + d_G(u))^2 \left[\frac{d_G(v)(d_G(v) - 1)}{2}\right] \\
&= 16m + 18Z_1(G) + Z_2(G) + \frac{7}{2}\sum_{v \in V(G)} (d_G(v))^3 \\
&\quad + \frac{1}{2}\sum_{v \in V(G)} (d_G(v))^4.
\end{aligned}$$

Corollary 5.3.2. *[40] Let $T_{n,k}$ be the tadpole graph, W_{n+1} be the wheel and L_n be the ladder. Then*

(i) $Z_1(T(S(T_{n,k}))) = 64(n + k) + 28$,

(ii) $Z_1(T(S(W_{n+1}))) = 147n + 8n^2 + n^3$,

(iii) $Z_1(T(S(L_n))) = 270n - 284$.

Theorem 5.3.3. *Let G be a graph with n vertices and m edges. Then*

(i) $Z_1(T_1(S(G))) = 24m + 4Z_1(G)$,

(ii) $Z_2(T_1(S(G))) = 16m + 12Z_1(G)$.

Proof. By referring Fig. 5.1 we see that the edge set $E(T_1(S(G))) = E_1 \cup E_2 \cup E_3$. Therefore by Observations 5.2.4 to 5.2.6 we have

$$
\begin{aligned}
(i)\ Z_1(T_1(S(G))) &= \sum_{uv \in E(T_1(S(G)))} \left[d_{T_1(S(G))}(u) + d_{T_1(S(G))}(v) \right] \\
&= \sum_{uu_e \in E_1} \left[d_{T_1(S(G))}(u) + d_{T_1(S(G))}(u_e) \right] \\
&\quad + \sum_{ue' \in E_2} \left[d_{T_1(S(G))}(u) + d_{T_1(S(G))}(e') \right] \\
&\quad + \sum_{u_e e' \in E_3} \left[d_{T_1(S(G))}(u_e) + d_{T_1(S(G))}(e') \right] \\
&= \sum_{uv \in E(G)} \left[2d_G(u) + 4 + 2d_G(v) + 4 \right] \\
&\quad + \sum_{uv \in E(G)} \left[2d_G(u) + 2 + 2d_G(v) + 2 \right] \\
&\quad + \sum_{uv \in E(G)} \left[4 + 2 + 4 + 2 \right]
\end{aligned}
$$

$$= [8m + 2Z_1(G)] + [4m + 2Z_1(G)] + 12m$$

$$= 24m + 4Z_1(G).$$

$$(ii) \ Z_2(T_1(S(G))) = \sum_{uv \in E(T_1(S(G)))} \left(d_{T_1(S(G))}(u) \right) \left(d_{T_1(S(G))}(v) \right)$$

$$= \sum_{uu_e \in E_1} \left(d_{T_1(S(G))}(u) \right) \left(d_{T_1(S(G))}(u_e) \right)$$

$$+ \sum_{ue' \in E_2} \left(d_{T_1(S(G))}(u) \right) \left(d_{T_1(S(G))}(e') \right)$$

$$+ \sum_{u_e e' \in E_3} \left(d_{T_1(S(G))}(u_e) \right) \left(d_{T_1(S(G))}(e') \right)$$

$$= \sum_{uv \in E(G)} [(2d_G(u))(4) + (2d_G(v))(4)]$$

$$+ \sum_{uv \in E(G)} [(2d_G(u))(2) + (2d_G(v))(2)]$$

$$+ \sum_{uv \in E(G)} [(4)(2) + (4)(2)]$$

$$= 16m + 12Z_1(G).$$

$\square$

Theorem 5.3.4. *Let G be a graph with n vertices, m edges and vertex set $V(G)$. Then*

$$(i) \ Z_1(T_2(S(G))) = 12m + 5Z_1(G) + \sum_{v \in V(G)} (d_G(v))^3,$$

$$(ii) \ Z_2(T_2(S(G))) = 8m + 6Z_1(G) + Z_2(G) + \frac{5}{2} \sum_{v \in V(G)} (d_G(v))^3$$

$$+ \frac{1}{2} \sum_{v \in V(G)} (d_G(v))^4.$$

Proof. By referring Fig. 5.1 we see that the edge set $E(T_2(S(G))) = E_2 \cup E_3 \cup E_4 \cup E_5$. Therefore by Observations 5.2.4 to 5.2.6 we have

$$
\begin{aligned}
(i)\ Z_1(T_2(S(G))) &= \sum_{uv \in E(T_2(S(G)))} \left[d_{T_2(S(G))}(u) + d_{T_2(S(G))}(v) \right] \\
&= \sum_{ue' \in E_2} \left[d_{T_2(S(G))}(u) + d_{T_2(S(G))}(e') \right] \\
&\quad + \sum_{u_e e' \in E_3} \left[d_{T_2(S(G))}(u_e) + d_{T_2(S(G))}(e') \right] \\
&\quad + \sum_{e'e'' \in E_4} \left[d_{T_2(S(G))}(e') + d_{T_2(S(G))}(e'') \right] \\
&\quad + \sum_{e''f' \in E_5} \left[d_{T_2(S(G))}(e'') + d_{T_2(S(G))}(f') \right] \\
&= \sum_{uv \in E(G)} \left[d_G(u) + 2 + d_G(u) + d_G(v) + 2 + d_G(v) \right] \\
&\quad + \sum_{uv \in E(G)} \left[2 + 2 + d_G(u) + 2 + 2 + d_G(v) \right] \\
&\quad + \sum_{uv \in E(G)} \left[2 + d_G(u) + 2 + d_G(v) \right] \\
&\quad + \sum_{e''f' \in E_5} \left[2 + d_G(v) + 2 + d_G(v) \right] \\
&= \left[4m + 2Z_1(G) \right] + \left[8m + Z_1(G) \right] + \left[4m + Z_1(G) \right] \\
&\quad + \sum_{v \in V(G)} (4 + 2d_G(v)) \left[\frac{d_G(v)(d_G(v) - 1)}{2} \right] \\
&= 12m + 5Z_1(G) + \sum_{v \in V(G)} (d_G(v))^3.
\end{aligned}
$$

$$
\begin{aligned}
(ii)\ Z_2(T_2(S(G))) &= \sum_{uv \in T_2(S(G))} \left(d_{T_2(S(G))}(u) \right) \left(d_{T_2(S(G))}(v) \right) \\
&= \sum_{ue' \in E_2} \left(d_{T_2(S(G))}(u) \right) \left(d_{T_2(S(G))}(e') \right)
\end{aligned}
$$

$$+ \sum_{u_e e' \in E_3} \left(d_{T_2(S(G))}(u_e)\right) \left(d_{T_2(S(G))}(e')\right)$$

$$+ \sum_{e' e'' \in E_4} \left(d_{T_2(S(G))}(e')\right) \left(d_{T_2(S(G))}(e'')\right)$$

$$+ \sum_{e'' f' \in E_5} \left(d_{T_2(S(G))}(e'')\right) \left(d_{T_2(S(G))}(f')\right)$$

$$= \sum_{uv \in E(G)} \left[d_G(u)\left(2 + d_G(u)\right) + d_G(v)\left(2 + d_G(v)\right)\right]$$

$$+ \sum_{uv \in E(G)} \left[2\left(2 + d_G(u)\right) + 2\left(2 + d_G(v)\right)\right]$$

$$+ \sum_{uv \in E(G)} \left(2 + d_G(u)\right)\left(2 + d_G(v)\right)$$

$$+ \sum_{e'' f' \in E_5} \left(2 + d_G(v)\right)\left(2 + d_G(v)\right)$$

$$= \sum_{uv \in E(G)} \left[2\left(d_G(u) + d_G(v)\right) + \left(\left(d_G(u)\right)^2 + \left(d_G(v)\right)^2\right)\right]$$

$$+ \sum_{uv \in E(G)} \left[8 + 2\left(d_G(u) + d_G(v)\right)\right]$$

$$+ \sum_{uv \in E(G)} \left[4 + 2\left(d_G(u) + d_G(v)\right) + \left(d_G(u)\right)\left(d_G(v)\right)\right]$$

$$+ \sum_{v \in V(G)} \left[4 + 4d_G(v) + \left(d_G(v)\right)^2\right] \left[\frac{d_G(v)\left(d_G(v) - 1\right)}{2}\right]$$

$$= \left[2Z_1(G) + \sum_{v \in V(G)} \left(d_G(v)\right)^3\right] + \left[8m + 2Z_1(G)\right]$$

$$+ \left[4m + 2Z_1(G) + Z_2(G)\right]$$

$$+ \left[2Z_1(G) - 4m + 2 \sum_{v \in V(G)} \left(d_G(v)\right)^3 - 2Z_(G)\right.$$

$$+\frac{1}{2} \sum_{v \in V(G)} (d_G(v))^4 - \frac{1}{2} \sum_{v \in V(G)} (d_G(v))^3 \Bigg]$$
$$= 8m + 6Z_1(G) + Z_2(G) + \frac{5}{2} \sum_{v \in V(G)} (d_G(v))^3$$
$$+\frac{1}{2} \sum_{v \in V(G)} (d_G(v))^4.$$

$\square$

5.4 Zagreb Co-indices

If G has n vertices and m edges, then

(i) $T(S(G))$ has $n + 3m$ vertices and $6m + \frac{1}{2}Z_1(G)$ edges.

(ii) $T_1(S(G))$ has $n + 3m$ vertices and $6m$ edges.

(iii) $T_2(S(G))$ has $n + 3m$ vertices and $4m + \frac{1}{2}Z_1(G)$ edges.

Gutman and Zhou [25] gave the following results.

Theorem 5.4.1. *[25] Let G be a graph with n vertices and m edges.*
Then

(i) $\overline{Z_1}(G) = 2m(n-1) - Z_1(G);$

(ii) $\overline{Z_2}(G) = 2m^2 - \frac{1}{2}Z_1(G) - Z_2(G).$

Knowing the number of vertices and edges of $T(S(G))$, $T_1(S(G))$ and of $T_2(S(G))$ and using Theorem 5.4.1 along with the results of Section 5.3 we get following.

Zagreb Indices and Co-indices of Total graph, Semi-total Point Graph and Semi-total Line Graph of Subdivision Graphs

Theorem 5.4.2. *Let G be a graph with n vertices, m edges and vertex set $V(G)$. Then*

$$(i) \;\; \overline{Z_1}(T(S(G))) \;=\; 12m(n + 3m - 3) + (n + 3m - 9)Z_1(G)$$
$$- \sum_{v \in V(G)} (d_G(v))^3.$$

$$(ii) \;\; \overline{Z_2}(T(S(G))) \;=\; m(72m - 28) + \left[12m - 22 + \frac{1}{2}Z_1(G)\right]Z_1(G)$$
$$- Z_2(G) - 4\sum_{v \in V(G)} (d_G(v))^3 - \frac{1}{2}\sum_{v \in V(G)} (d_G(v))^4.$$

$$(iii) \;\; \overline{Z_1}(T_1(S(G))) \;=\; 12m(n + 3m - 3) - 4Z_1(G).$$

$$(iv) \;\; \overline{Z_2}(T_1(S(G))) \;=\; 72m^2 - 28m - 14Z_1(G).$$

$$(v) \;\; \overline{Z_1}(T_2(S(G))) \;=\; 4m(2n + 6m - 5) + (n + 3m - 6)Z_1(G)$$
$$- \sum_{v \in V(G)} (d_G(v))^3.$$

$$(vi) \;\; \overline{Z_2}(T_2(S(G))) \;=\; m(32m - 14) + \left[8m + \frac{1}{2}Z_1(G) - \frac{17}{2}\right]Z_1(G)$$
$$- Z_2(G) - 3\sum_{v \in V(G)} (d_G(v))^3$$
$$- \frac{1}{2}\sum_{v \in V(G)} (d_G(v))^4.$$

Chapter 6

Friendship Network Analysis Using Status Connectivity Indices of Graphs

6.1 Introduction

Friendship is a common place of notion, familiar to and cherished by people around the world and across history. The capacity for friendship is as fundamental to the human condition as are familial attachment; romantic, conjugal, and sexual loving; competition; and conflict. Since the time of Aristotle, friendship has been recognized as essential to the well-lived life. However, only in the last 35 years has markedly increased social scientific empirical work emerged addressing friendship as a distinctive category of human experience, cognitive and moral development, personal and social relationship, and communicative, associational, and political activity. Despite and perhaps because of its pervasive presence in human life, the word "friendship" references a vexing, intellectually captivating, continually evolving, and diversely apprehended array of phenomena. The words "friend" and "friendship" are used to describe a gamut of human relationships, ranging from long-standing attachments of considerable affection and loyalty, to someone just met at some place. Further complicating matters, friendship is unique in its capacity to arise as a free-standing relationship on its own terms between two persons, or as a sincerely lived dimension of other relationships, such as the friendship developed between siblings, spouses, parents and children, or coworkers. It is a negotiated attachment between persons that always reflects shared personal dispositions and material sociocultural possibilities. Friend-ships are important, potentially lifelong, close relationships that are essential to our social, psychological, and physical well- being [55].

Finally, friends play a key role in providing intimacy, as the level of disclosure tends to be more intense with close friends than with other peers.

Methods to estimate the degree rank of a node, that are faster than the classical method of computing the centrality value of all nodes and then rank a node is proposed and proposed methods are modeled based on the network characteristic and sampling techniques in [57] and the pattern of personal sharing between male and female students in urban and rural area students reported in [66].

In this chapter the concept of connectivity and friendship be-tween individuals when they are brought together at one place from different cultural background and environments are studied by using mathematical parameters namely status connectivity indices to ana-lyze the basic relations among the members of peer group coming from different places as they spend time together in different curricular and co-curricular activities.

6.2 Method

The participants involved in this study are M.Sc. first year students of Mathematics at Karnatak University, Dharwad admitted during academic year 2019-20.

We use the graph model with topological indices to analyze the information collected from the students.

The *status* of a vertex $u \in V(G)$, denoted by $\sigma_G(u)$ is defined as [28],

$$\sigma_G(u) = \sum_{v \in V(G)} d_G(u, v).$$

The *first status connectivity index* $S_1(G)$ and *second status connectivity index* $S_2(G)$ of a graph G are defined as [51]

$$S_1(G) = \sum_{uv \in E(G)} [\sigma_G(u) + \sigma_G(v)] \quad \text{and} \quad S_2(G) = \sum_{uv \in E(G)} \sigma_G(u)\sigma_G(v). \tag{6.1}$$

As the degree of a vertex increases, the status decreases. Thus as degree increases, that is closed neighbors increases, the number of edges in a graph increases and in turn the status connectivity indices increases as the summation in Eq. (6.1) runs over all edges of a graph.

6.3 Results and Discussion

Information from the students after their admission to the M.Sc. course in the form of questionnaire whether they have contacted other students or talked with them about certain matters or to build friendships or for help or to get any other information was collected. Around 70 students responded out of 82. The group consists of both boys and girls. But girls are more around 80%. We collected the information in two phases. First phase was conducted at the beginning of the semester (after 15 days from the starting of course) as they were fresh to new environment and most of them were unknown to university campus life. The second phase conducted after three months from the first phase.

The information gathered is modeled as graphs G_1 and G_2, depicted in Figs. 6.1 and 6.2. The vertices represents the students labeled with the numbers 1 to 70. Two vertices are joind by an edge if the corresponding students build the relation among themselves by contacting at lest once during the phase.

The status and degrees of the vertices in graphs of both phases are listed in Table 6.1 (vertices labeled from 1 to 36) and Table 6.2 (vertices labeled from 37 to 70). Some of the vertices has status zero in G_1. This indicates that the corresponding student does not have any contact, so no friend at the beginning of the semester. In graph

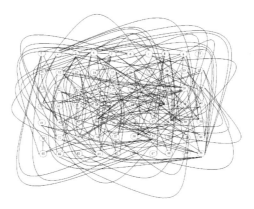

Figure 6.1 : Graph G_1 based on data collected for first time.

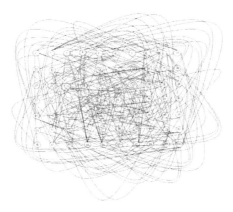

Figure 6.2 : Graph G_2 based on data collected for second time.

G_2, the status of each point is more than zero. It indicates that as time goes, almost every one has at least one friend in the group. By comparing the data of two graphs as given in Table 6.1 (vertices labeled from 1 to 36) and table 6.2 (vertices labeled from 37 to 70), we observe that the degrees of the vertices are increasing in second graph compared to first graph, whereas the status of the vertices decreases in second graph compared with first graph. It indicates that as students spend more time in a class, they increases the contacts with others. That is, friendship increases. Further we observe that as degree increases, the number of connections (friends) increases. Hence the status connectivity indices increases which can be seen from the Table 6.3, as the status connectivity indices has more value in graph G_2 than in graph G_1. This shows that as time span increases, the number of friends increases.

6.4   Conclusion

The study of friendship between the individuals of newly admitted students, who came from different places, is considered at two phases. First phase is at the begining of the semester and second phase after three months gap from first phase is carried out. Using graph model, based on the information gathered, we analysed the study of friendship with the help of topological indices namely status connectivity indices. As the students spends more time in the campus, the interaction between them increases and more new friends will be connected.

Friendship Network Analysis Using Status Connectivity Indices of Graphs

Table 6.1: Status and degrees of vertices in graphs G_1 and G_2

Vertex	Status of vertex in G_1	degree of vertex in G_1	Status of vertex in G_2	degree of vertex in G_2
1	221	5	169	7
2	240	6	181	9
3	164	12	160	13
4	133	16	130	17
5	168	10	145	13
6	192	5	189	6
7	151	10	149	11
8	138	12	130	13
9	135	12	132	14
10	0	0	168	6
11	158	5	154	6
12	165	9	190	10
13	202	6	197	8
14	249	2	245	3
15	232	5	213	5
16	168	7	165	9
17	0	0	257	1
18	221	8	146	15
19	233	4	163	10
20	191	6	168	6
21	0	0	235	1
22	220	7	217	9
23	184	5	154	10
24	174	5	159	10
25	174	6	170	7
26	0	0	182	4
27	0	0	186	3
28	160	10	158	11
29	148	14	140	15
30	220	5	194	6
31	209	4	163	8
32	215	4	199	5
33	157	12	153	13
34	156	10	156	11
35	210	6	190	8
36	190	3	181	4

145

Table 6.2: Status and degrees of vertices in graphs G_1 and G_2 (Continued from table 6.1.)

Vertex	Status of vertex in G_1	degree of vertex in G_1	Status of vertex in G_2	degree of vertex in G_2
37	0	0	236	2
38	164	11	157	12
39	191	6	175	8
40	173	6	168	7
41	228	4	177	7
42	176	7	171	8
43	187	7	185	8
44	164	13	160	14
45	214	5	194	6
46	0	0	189	3
47	144	9	137	14
48	173	13	153	15
49	173	7	169	9
50	160	7	158	7
51	235	1	196	5
52	175	7	173	9
53	172	6	169	10
54	191	7	188	9
55	169	6	163	8
56	164	6	160	7
57	184	9	173	11
58	235	1	178	7
59	187	4	143	11
60	164	6	155	9
61	180	4	175	7
62	184	5	154	11
63	234	1	186	3
64	165	11	157	13
65	202	3	143	13
66	236	2	160	10
67	200	1	138	14
68	176	4	175	5
69	222	1	218	4
70	135	14	128	15

Table 6.3: Status indices of graphs

Index	Graph G_1	Graph G_2
S_1	73552	98803
S_2	6411448	8054229

References

[1] G. Al Hagri, M. El Marraki, E. Essalih, The degree distance of certain particular graphs, Appl. Math. Sci., **6** (2012) 857–867.

[2] P. Ali, S. Murkwembi, S. Munyira, Degree distance and vertex connectivity, Discrete Appl. Math. **161** (2013) 2802–2811.

[3] M. Aouchiche, P. Hansen, Distance spectra of graphs: A survey, Linear Algebra Appl. **458** (2014) 301–386.

[4] A. R. Ashrafi, T. Došlić, A. Hamzeh, Extremal graphs with respect to the Zagreb co-indices, MATCH Commun. Math. Compu. Chem., **65** (2011) 85–92.

[5] A. R. Ashrafi, M. Saheli, M. Ghorbani, The eccentric connectivity index of nanotubes and nanotori, J. Comput. Appl. Math., **235** (2011) 4561–4566.

[6] O. Bucicovschi, S. M. Cioabă, The minimum degree distance of a graphs of given order and size, Discrete Appl. Math., **156** (2008) 3518–3521.

[7] R. Chang, Y. Zhu, On the harmonic index and the minimum degree of a graph, Romanian J. Inf. Sci. Tech., **15** (2012) 335–343.

[8] P. Dankelmann, I. Gutman, S. Murkwembi, H. C. Swart, On the degree distance of a graph, Discrete Appl. Math., **157** (2009) 2773–2777.

[9] K. C. Das, D. Lee, A. Graovac, Some properties of the Zagreb eccentricity indices, Ars Math. Contem., **6** (2013) 117–125.

[10] K. C. Das, G. Su, L. Xiong, Relation between degree distance and Gutman index of graphs, MATCH Commun. Math. Comput. Chem., **76** (2016) 221–232.

[11] H. Deng, S. Balachandran, S. K. Ayyaswamy, Y. B. Venkatakrishnan, On the harmonic index and the chromatic number of a graph, Discrete Appl. Math., **161** (2013) 2740–2744.

[12] J. Devillers, A. T. Balaban (Eds.), Topological Indices and Related Descriptors in QSAR and QSPR, Gordon & Breach, Amsterdam, 1999.

[13] A. A. Dobrynin, R. Entringer, I. Gutman, Wiener index of trees: theory and applications, Acta Appl. Math., **66** (2001) 211–249.

[14] A. A. Dobrynin, A. A. Kochetova, Degree distance of a graph: a degree analogue of the Wiener index, J. Chem. Inf. Comput. Sci., **34** (1994) 1082–1086.

[15] T. Došlić, M. Saheli, Eccentric connectivity index of benzenoid graphs, in: I. Gutman, B. Furtula (Eds.), Novel Molecular Structure Descriptors-Theory and Applications II, Uni. Kragujevac, Kragujevac, 2010, 169–183.

[16] B. W. Douglas, Introduction to Graph Theory, Prentice Hall, New Delhi, 2001.

[17] Z. Du, B. Zhou, Degree distance of unicyclic graphs, Filomat, **24** (2010) 95–120.

[18] E. Estrada, L. Torres, L. Rogríguez, I. Gutman, An atom-bond connectivity index: Modelling the enthalpy of formation of alkanes, Indian J. Chem., **37A** (1998) 849–855.

[19] S. Fajtlowicz , On conjectures of Graffiti-II, Congr. Number, **60** (1987) 187–197.

[20] B. Furtula, A. Graovac, D. Vukicević, Augmented Zagreb index, J. Math. Chem., **48** (2010) 370–380.

[21] B. Furtula, I. Gutman, A forgotten topological index, J. Math. Chem., **53** (2015) 1184–1190.

[22] S. Gupta, M. Singh, A. K. Madan, Application of graph theory: relationship of eccentric connectivity index and Wiener's index with anti-inflammatory activity, J. Math. Anal. Appl., **266** (2002) 259–268.

[23] I. Gutman, Selected properties of the Schultz molecular topological index, J. Chem. Inf. Comput. Sci., **34** (1994) 1087–1089.

[24] I. Gutman, Degree-based topological indices, Croat. Chem. Acta, **86** (2013) 351–361.

[25] I. Gutman, B. Furtula, Z. K. Vukecevic and G. Popivoda, On Zagreb indices and Coindices, MATCH Commun. Math. Comput. Chem., **74** (2015) 5-16.

[26] I. Gutman, N. Trinajstić, Graph theory and molecular orbitals. Total π-electron energy of alternant hydrocarbons, Chem. Phys. Lett., **17** (1972) 535–538.

[27] I. Gutman, Y. Yeh, S. Lee, Y. Luo, Some recent results in the theory of the Wiener number, Indian J. Chem., **32A** (1993) 651–661.

[28] F. Harary, Status and contrastatus, Sociometry, **22** (1959) 23–43.

[29] F. Harary, Graph Theory, Narosa Publishing House, New Delhi, 1999.

[30] Y. Hu and X. Zhou, On the harmonic index of the unicyclic and bicyclic graphs, WSEAS Tran. Math., **12** (2013) 716–726.

[31] H. Hua, K. C. Das, The relationship between the eccentric connectivity index and Zagreb indices, Discrete Appl. Math., **161** (2013) 2480–2491.

[32] A. Ilić, S. Klavžar, D. Stevanović, Calculating the degree distance of partial hamming graphs, J. Chem. Inf. Comput. Sci., **1** (2009) 1–14.

[33] A. Ilić, D. Stevanović, L. Feng, G. Yu, P. Denkelmann, Degree distance of unicyclic and bicyclic graphs, Discrete Appl. Math., **159** (2011) 779–788.

[34] O. Ivanciuc, T. S. Balaban, A. T. Balaban, Design of topological indices, Part 4, Reciprocal distance matrix, related local vertex invariants and topological indices, J. Math. Chem., **12** (1993) 309–318.

[35] S. Kanwal, I. Tomescu, Bounds for degree distance of a graph, Math. Reports, **17** (2015) 337–344.

[36] V. Kumar, S. Sardana, A. K. Madan, Predicting anti-HIV activity of 2, 3-diaryl-1, 3 thiazolidin-4-ones: computational approach using reformed eccentric connectivity index, J. Mol. Model., **10** (2004) 399–407.

[37] J. Liu, On the harmonic index of triangle free graphs, Appl. Math., **4** (2013) 1204–1206.

[38] J. Liu , On the harmonic index and diameter of graphs, Appl. Math. Phy., **1** (2013) 5–6.

[39] R. Mohammadyari, M. R. Darafsheh, Topological indices of the Kneser graph $KG_{n,k}$, Filomat, **26** (2012) 665–672.

[40] G. Mohanappriya , D. Vijayalakshmi, Topological indices of total graph of subdivision graphs, Annals Pure Appl. Math., **14** (2017) 231–235.

[41] M. J. Morgan, S. Mukwembi, H. C. Swart, On the eccentric connectivity index of a graph, Discrete Math., **311** (2011) 1229–1234.

[42] K. P. Narayankar, D. Selvan, Geometric arithmetic status index of graphs, Int. J. Math. Arch., **8** (2017) 230–233.

[43] S. Nikolić, G. Kovačević, A. Miličević, N. Trinajstić, The Zgreb indices 30 years after, Croat. Chem. Acta, **76** (2003) 113–124.

[44] S. Nikolić, A. Miličević, N. Trinajstić, A. Jurić, On use of the variable Zagreb $^{v}M_2$ index in QSPR: Boiling points of benzenoid hydrocarbons, Molecules, **9** (2004) 1208–1221.

[45] D. Plavšić, S. Nikolić, N. Trinajstić, Z. Mihalić, On the Harary index for the characterization of chemical graphs, J. Math. Chem., **12** (1993) 235–250.

[46] H. S. Ramane, B. Basavanagoud, A. S. Yalnaik, Harmonic status index of graphs, Bull. Math. Sci. Appl., **17** (2016) 24–32.

[47] H. S. Ramane, I. Gutman, A. B. Ganagi, On diameter of line graphs, Iranian J. Math. Sci. Inf., **8** (2013) 105–109.

[48] H. S. Ramane, V. V. Manjalapur, I. Gutman, General sum-connectivity index, general product-connectivity index, general Zagreb index and co-indices of line graph of subdivision graphs, AKCE Int. J. Graphs Combin., **14** (2017) 92–100.

[49] H. S. Ramane, D. S. Revankar, A. B. Ganagi., On the Wiener index of a graph, J. Indones. Math. Soc., **18** (2012) 57–66.

[50] H. S. Ramane, D. S. Revankar, I. Gutman, H. B. Walikar, Distance spectra and distance energies of iterated line graphs of regular graphs, Publ. Inst. Math. Beograd, **85** (2009) 39–46.

[51] H. S. Ramane, A. S. Yalnaik, Status connectivity indices of graphs and its applications to the boiling point of benzenoid hydrocarbons, J. Appl. Math. Comput., **55** (2017) 609–627.

[52] H. S. Ramane, A. S. Yalnaik, Bounds for the status connectivity index of line graphs, Int. J. Comput. Appl. Math., **12** (2017) 305–310.

[53] H. S. Ramane, A. S. Yalnaik, R. Sharafdini, Status connectivity indices and co-indices of graphs and its computation to some

distance-balanced graphs, AKCE Int. J. Graphs Combin., (2018) (in press) https://doi.org/10.1016/j.akcej.2018.09.002.

[54] P. S. Ranjini, V. Lokesha, I. N. Cangül, On the Zagreb indices of the line graphs of the subdivision graphs, Appl. Math. Comput., **218** (2011) 699–702.

[55] B. R. Rawlins, Measuring the relationship between organizational transparency and employee trust, Public Relations Journal, **2** (2009) 1-21.

[56] E. Sampathkumar, S. B. Chikkodimath, Semitotal graphs of a graph - I, J. Karnatak Univ. Sci., **18** (1973) 274–280.

[57] A. Saxena, R. Gera, S. R. S. Iyengar, Estimating degree rank in complex networks, Social Network Analysis and Mining, **8** (42) (2018) pages 20.

[58] R. Sharafdini, T. Reti, On the transmission-based graph topological indices, Kragujevac J. Math., **44** (2020) 41–63.

[59] V. Sheeba Anges, Degree distance and Gutman index of corona product of graphs, Trans. Comb., **4** (2015) 11–23.

[60] V. S. Shigehalli, R. Kanabur, Arithmetic-geometric indices of path graph, J. Comput. Math. Sci., **6** (2015) 19–24.

[61] B. S. Shetty , V. Lokesha, P. S. Ranjini, On the harmonic index of graph operations, Trans. Combin., **4** (2015) 5–14.

[62] R. Todeschini, V. Consonni, Handbook of Molecular Descriptors, Wiley-VCH, Weinheim, 2000.

[63] I. Tomescu, Some extremal properties of the degree distance of a graph, Discrete Appl. Math., **98** (1999) 159–163.

[64] A. I. Tomescu, Unicyclic and bicyclic graphs having minimum degree distance, Discrete Appl. Math., **156** (2008) 125–130.

[65] I. Tomescu, Properties of connected graphs having minimum degree distance, Discrete Math., **309** (2009) 2745–2748.

[66] B. Vasanthi, S. Arumugam, S. V. Nayana, Personal sharing in small groups - A case study, Proc. Social Beha. Sci., **219** (2016) 108–112.

[67] D. Vukicević, B. Furtula, Topological index based on the ratios of geometrical and arithmetical means of end-vertex degrees, J. Math. Chem., **46** (2009) 1369–1376.

[68] D. Vukičević, A. Graovac, Note on the comparison of the first and second normalized Zagreb eccentricity indices, Acta Chim. Slov., **57** (2010) 524–528.

[69] D. B. West, Introduction to Graph Theory, Prentice Hall of India, New Delhi, 2000.

[70] H. Wiener, Structural determination of paraffin boiling points, J. Am. Chem. Soc., **69** (1947) 17–20.

[71] K. Xu, K. C. Das, N. Trinajstić, The Harary Index of a Graph, Springer, Heidelberg, 2015.

[72] L. Zhong, The harmonic index of graphs, Appl. Math. Lett., **25** (2012) 561–566.

[73] B. Zhou, Z. Du, On eccentric connectivity index, MATCH Commun. Math. Comput. Chem., **63** (2010) 181–198.

[74] B. Zhou, N. Trinajstić, On a novel connectivity index, J. Math. Chem., **46** (2009) 1252–1270.

[75] Y. Zhu, R. Chang, On the harmonic index of bicyclic conjugated molecular graphs, Filomat, **28** (2014) 421–428.

CPSIA information can be obtained
at www.ICGtesting.com
Printed in the USA
BVHW051757250523
664860BV00016BA/602